A HISTORY OF GLASSMAKING

A HISTORY OF GLASSMAKING

R. W. DOUGLAS
and
SUSAN FRANK

G T FOULIS & CO LTD
Henley-on-Thames, Oxfordshire

ISBN 0 85429 117 2

Printed in Great Britain by
The Whitefriars Press Ltd., London and Tonbridge

Contents

Preface

Our aim in writing this book was essentially to provide a convenient summary of the history of glassmaking for the many different kinds of people who have some interest in glass. The study of glass has various aspects, some of which interest the artist, some the engineer and others the scientist or the industrialist.

Perhaps because glass possesses a visual history in the articles which have been preserved through the centuries, perhaps because of its unique properties as a material, people of many different interests find in their appreciation of glass something in common. In endeavouring to write a story of the growth of an industry and an art one has to refer to the details of technical processes and to the advances of science which from time to time contribute to the development of those processes. We hope that we have had some success in making the story clear and interesting to the layman and the expert.

It was a historian who said that to write a history of mathematics one must know some mathematics and, as it will be found, we have not discovered how to write the history of a technology without introducing some technology. Perhaps even more difficult has been the necessity of attempting to deal with the scientific background to the technology. Here we shall have risked writing in too obvious terms for the technologist and still remaining incomprehensible to the non-scientist.

We have endeavoured to trace from the earliest times until the present the development of glass technology as it appears to us. Although at times we may appear to be concentrating too much upon the development of the glass industry in Britain, we have endeavoured to deal with the changes which have occurred in other countries when it appeared that the main impetus had for the time being shifted to those countries.

It will be obvious to the technologist that this is not a textbook of technology but rather an attempt to take the broad view and see the technology developing against, and interacting with, a background of scientific and social change.

In keeping the book to a reasonable length we have at times had to make quite arbitrary choices of the amount of detail in which any one topic is discussed. Perhaps we should also deny any attempt to be writing history for historians; we realize that many people are devoting the whole of their energies to studying the history of science while others are concerned with the history of the industrial arts.

The historians of science at first tended to concentrate upon the history of ideas; in writing this book it is the quantitative methods in experimental science that we have stressed as major factors in the development of the technology. That we should have done this will not surprise many people who have practised science for a long period and there is support for this view to be found in the literature. The following quotation from Isaac Newton is but one example:

> For the best and safest method of philosophizing seems to be, first diligently to investigate the properties of things and establish them by experiment, and then to seek hypotheses to explain them. For hypotheses ought to be fitted merely to explain the properties of things and not to attempt to predetermine them except in so far as they can be an aid to experiments.

We have drawn up many sources and these are acknowledged in the bibliography given at the end of each chapter. Many of the secondary sources are fairly readily available and we hope we shall have aroused sufficient interest for readers to turn to some of these for themselves.

A historian when discussing historical thinking said:

> But now we think of things as in a ceaseless flux; and though that makes them much more difficult and complex to grasp, we are at any rate nearer to understanding them, or at least describing them as they are.

We have been describing only one small part of that ceaseless flux but in doing so we have felt more and more strongly the need to emphasize the continuous contribution from all manner of sources to that small area.

R. W. Douglas
Susan Frank
Sheffield, 1972

'Is there any thing whereof it may be said, See, this is new?
it hath been already of old time, which was before us.'

Ecclesiastes, i, 10

1
The history of glassmaking from earliest times to the nineteenth century

Life without glass is difficult to imagine; glass windows allow light from the sun to brighten our buildings, but they can also help to keep our rooms warm in winter; glass is essential for the provision of convenient artificial light, and as a component in the cameras, cine-projectors, and television sets which we use during our leisure time. It provides us with fine drinking glasses and robust cooking utensils, and with containers for food and liquids of all kinds. This hard, inert, transparent material is made by heating together a mixture of materials such as sand, limestone and soda. At a sufficiently high temperature, a white heat of about 1400-1500°C just below the melting point of iron, these materials react to form a liquid. When this liquid is taken from the furnace it gets stiffer and stiffer as it cools until at about 500°C it has become as solid as the glass we are familiar with in our windows or on our tables. In more scientific terms, glasses are made by heating a mixture of materials which react with one another to form a quiescent melt; this melt is a viscous liquid which cools with increasing viscosity until it becomes a rigid solid. If the cooling is too rapid, the glass does not have time to release the stresses set up within it during cooling and it is liable to shatter when it becomes a solid. For this reason glass articles are subjected to a controlled heat treatment after manufacture which releases the stresses in the glass, a process known as annealing.

The ingredients of the most common commercial glasses are sand, limestone and soda: the soda acts as a flux and the limestone as a stabilizing agent to give a durable glass. Glasses are really supersaturated solutions of the fluxes in silica and as such would crystallize, if they were able to reach their equilibrium condition. The skill in glassmaking lies in choosing mixtures in which crystallization, or devitrification, is so extremely slow that in practice it is not observed.

Glasses can be made of an infinite variety of compositions. A large number of important types of glass have been discovered following the developments of chemical science—special glasses for lens making; a glass containing boric oxide as a flux, replacing most of the soda, is used for oven ware; and new glasses containing no soda at all are used for mercury vapour lamps for street lighting or for many fibre glass applications.

This book will be concerned with the history of glassmaking and the influence upon the technology of agents ranging from excise duties to the discoveries of modern science, from the demands of military strategy to the pressures of commerce and politics. The major part of the book will be concerned with the last two hundred years and particularly with the last eighty years during which time modern science has flowered and revolutionized the technology. Nevertheless, the development has been continuous from the earliest times and this first chapter will survey the history briefly until about 1800 AD.

The first glassmakers

The vitreous glaze on pottery could be described as a glass, and it was in use long before glass was employed as an independent substance. It is thought that glassmaking originated in western Asia around the third millennium BC and that the technique was later brought to Egypt; originally only beads and similar small objects were made but glass vessels appeared around 1500 BC. Fragments of glass vessels have been found on western Asiatic sites in strata of the late sixteenth and early fifteenth centuries BC, and in Egypt vessels were first made during the reign of Thutmose III (1504-1450 BC). Thutmose began a series of Asiatic conquests in 1481 BC, and it may be that he brought back workers to set up a glass-vessel industry in Egypt. The decoration on all Egyptian glass of about 1300 BC onwards is very similar to that found on glass from Ur, dating from about 1300 BC. In fact the extensive use of cobalt in Egyptian glass suggests that they derived their knowledge from Mesopotamia, as the nearest sources of cobalt were in Iran.

Glassmaking flourished in Mesopotamia and Egypt until about 1200 BC when peoples from Libya and Asia started to threaten the Egyptian state. These troubles coincided with internal decay and conditions were too unsettled for industry to flourish, although beads and small ornaments continued to be made after this date. The Syrians were especially important in keeping glassmaking alive, and Syria and Mesopotamia became the two main centres of glass manufacture when revival started during the ninth century BC. Although the products of these regions were probably distinct, each catering for specific groups of customers, it is likely that both industries were under the influence of Phoenicians and were probably staffed by them. The Phoenicians were Syrians who lived along the coast of present-day Lebanon: they gained a living from trading overseas, spreading the products of the glassmakers, amongst other things, throughout the ancient world. Glassmaking centres grew up in regions further west including Cyprus, Rhodes and probably Greece. The glass industry of the Italian peninsula dates from at least the ninth century BC, and in the fifth century BC either this industry itself or its products spread to the region around modern Venice and up into what is now Austria, around the village of Hall.

After the conquests of Alexander the Great towards the end of the fourth century BC, glassmaking in Mesopotamia declined, but the Syrian workshops prospered, their speciality being plain, moulded, monochrome bowls in a variety of colours. For the first time since the eleventh century BC, with the building of the glassworks in Alexandria, the city founded by Alexander the Great in 332 BC, Egypt again became known as a glassmaking centre. Alexandria attracted workers from a wide region and exported fine glassware as far as Greece and Italy. The Hellenistic techniques were probably brought to the Italian peninsula by the Alexandrians in the first century BC, and the Syrians also established glassworks in northern Italy around the beginning of the Christian era.

The invention of glass blowing

The discovery of glass blowing, the first great revolution in glassmaking, occurred in about 50 BC, probably in Phoenicia. The importance of this discovery will be better appreciated after the earlier processes have been described.

Throughout the eastern Mediterranean area, prior to the Christian era, glass vessels were made by starting with a shaped core probably made of mud bound with straw fixed to a metal rod. This core was covered with glass either by dipping the core into the molten glass or by winding molten threads of glass around it. The surface was then smoothed by continual reheating and rolling on a flat stone slab. Trailed or blobbed decoration was added by applying coloured threads which were pressed into the surface by rolling. In the trailing process, a thin tape or ribbon of glass was drawn out and wrapped, or trailed, around the glass surface rather in the way that decoration is added when icing a cake. Handles and footstands were then put on and after the core had been chipped out the vessel could be completed by the addition of a rim which was again trailed onto the glass. The major portion of pre-Roman glasses still in existence consists of such vessels.

Another method, known as the millefiori (thousand flowers) process, was also used to make open beakers or shallow dishes. A core was made of the shape of the inside of the required vessel and sections of monochrome or polychrome glass rods were laid out on the core and loosely fixed together with some adhesive material. An outer mould was placed in position to keep the sections together whilst the glass was fused. The core and outer mould were then removed and the surfaces of the vessel ground smooth to produce a striking mosaic effect—the glass surfaces displaying the cross-sections of the coloured rods. Such vials, unguent jars and perfume flasks were widely used by the ancient Greeks and Romans, and the history of the millefiori process is closely associated with that of the cosmetics industry. The process was developed in western Asia and may have been introduced by Asiatic workers to Alexandria where the glassworks became famous for their mosaic products.

Other manufacturing methods were similar to those used by the stone mason and the potter. Glass can be cut or ground by harder materials such as quartz and raw lumps of glass were cut and ground into pots or ornaments.

Bowl shapes were also made by pressing the glass into a mould of fired clay. In the lost wax process a wax replica of the required glass object was coated with clay which, having some strength whilst still unfired, held the shape when subsequently warmed to melt out the wax; the clay was then fired to form a mould into which the glass could be poured. Open vessels could also be made by fusing powdered glass *in situ* between two parts of a closed mould. Solid objects which were shaped on one side only such as flat-backed figurines were pressed in open moulds, whilst those which were modelled in the round were cast in two-piece moulds.

The processes described in the previous paragraphs were all in use before 1500 BC. Near to the beginning of the Christian era, men discovered how to blow glass. It may be that in order to make their work easier they lightened the iron rod used to gather the glass by making it hollow and then discovered that by blowing down this pipe they could form a hollow glass bubble. The gradual hardening of the molten glass as it cools made possible free, or off-hand, blowing to form a wide variety of shapes. The 'balloon' of glass could also be blown inside a mould; simple dishes and wide mouth containers could be blown in open moulds but soon two- or three-part moulds of wood or clay were developed, probably by the Syrians, and a much greater variety of shapes and designs could be produced more cheaply and more quickly. Transparent vessels could now be produced in contrast to the translucent or opaque articles made by the earlier methods. Plate 1 (between pages 36 and 37) shows a core-formed oinochoe or jug said to date from the fifth century BC, found in Rhodes; in contrast plate 3 shows a blown handled flagon found in Kent, probably dating from the late second or early third century AD.

The introduction of glass blowing gave a great impetus to the industry and within a century the art had spread via Persia to the Orient. In the West, the existence of the Roman Empire aided the foundation of many glassmaking centres; Syrian, Jewish and Alexandrian glass-blowers worked in Rome and the Saone and Rhine provinces. From there the art spread to Spain, the Low Countries, Gaul and Britain. The glassmakers could now produce, in quantity, simple blown table ware, as well as fine art glass; prices fell and glass could be purchased by an ever widening section of the population. The glass workshops of Rome became so numerous and the smoke from their furnaces caused such a nuisance that from 200 AD the city authorities forced glassmakers to concentrate in the suburbs away from the city.

Glass-workers in Egypt did not adopt the technique of blowing until some time in the second century AD, their specialities having been the making of mosaic and other fine coloured articles and the cutting of glass to produce cameos. This carving was a reflexion of the art which was highly developed throughout their history of carving in relief on stone to decorate their tombs. The famous Portland Vase, dating from around the beginning of the Christian era, combines the art of blowing with Egyptian skill in cameo cutting. Plate 4 (between pages 36 and 37) shows the vase, a dark cobalt blue inner layer of glass with an outer white casing of glass in which the design has been cut.

It is uncertain how the composite body of the vase was formed. It may be that the inner blue glass was gathered first and covered with the opaque white glass and the vase was then blown, but in 1873 John Northwood of Stourbridge who was famous for his glass cameo work began to make a replica of the Portland Vase using a technique which gave an outer layer with a much more even thickness than that of the original. This technique was known as cupping; a cup-shape of opaque white glass was made and whilst hot was completely filled with molten coloured glass gathered on the end of a blowing iron from the furnace, and the whole marvered to weld the two glasses together. Marvering is the process of shaping the glass on the end of the iron by rolling on a flat surface of stone or iron. Marble was once used for the stone and the name derives from the French *marbre*. The glass was then reheated, blown and shaped to give a glass blank, in the required size and shape, with the lower portion encased in opaque white glass; the handles were then added and the vase was ready for carving.

The carving of the Northwood vase was almost completed when it suddenly cracked, probably owing to strains in the two layers of glass. It is an extremely difficult technical problem to produce two glasses, such as the blue and opaque white, with sufficiently well matched physical properties, especially thermal expansion, so that they can be fused together at high temperature and cooled to give a strain-free object at room temperature. The ancient glass-blowers must have had many frustrating trials before the masterpiece was successfully made and such difficulties may account for so few surviving examples of these ancient skills.

The growth of glassmaking in Europe

The northern regions of the Roman Empire also had their glassmaking centres. Famous factories were established at Cologne and Trier which prospered for several centuries. It was at Trier that the late-Latin term 'glesum' originated, probably from a Germanic word for a transparent, lustrous substance. To this we owe our word 'glass'.

There was continual movement of glass-workers within the Roman Empire and it is often difficult and sometimes impossible to tell where a particular piece was made. Glass became so much a part of everyday life that glassware upon a Roman table was a sign of lack of affluence, the rich using gold and silver plate. The establishment of many widely dispersed centres during this period aided the continuation of the glass industry in Europe when the Roman Empire started to decline during the first half of the fifth century AD. Glass-blowers still worked in the cities of the Rhine and the Rhône, but many fled to Italy, particularly to the Po valley and to Altare (near Genoa) from where they later spread all over Europe. Techniques changed very little until the eleventh century AD: all glass of this period was blown and, like Roman glass, was of an alkali-lime-silica composition.

Although glasses were made by the Egyptians and the Romans having essentially the same compositions as the modern soda-lime-silica glasses used for making bottles and windows, the raw materials were not so well defined as today; a sand which contained the remains of shells probably provided the silica and the lime, while the soda which was used as a flux to form a melt with the sand would be present in the ash of sea plants.

The western European glassmakers were able to obtain the soda-rich marine-plant ash from the Mediterranean countries during the period following the collapse of the Roman Empire. Towards the end of the tenth century AD they began to use on a large scale ash from bracken and other woodland plants which had a high potash content and which could be easily obtained by glassmakers as they travelled about the country. At that time the woodlands were far more extensive than they are now and glassworks were established throughout wooded regions such as the Vosges and Bohemia. After the tenth century this potash glass became characteristic of central Europe whilst soda glass continued to be made in the coastal regions. During this period changes occurred in the style and decoration of glass articles. The art of making high-quality glass objects was generally lost and the only Roman techniques which survived were trailing and mould blowing. The customers demanded simpler, more primitive designs; thus most of the earlier techniques fell into disuse. The Church played an important part in keeping the glass industry alive, guarding the old glassmaking knowledge and acting as the patron of new and extensive applications of glass in glass mosaics and stained glass windows. The Church contained most of the people who could read and write and possessed the only libraries of the period in which the old glassmaking texts were kept; oral instructions for glass manufacture were written down by the clerics.

In the East also the industry prospered even after the Islamic conquests, and the tradition of painting, enamelling and gilding of glass is illustrated by the magnificent mosque lamps (see plate 5 between pages 36 and 37), many of which have survived in their original condition. The Mongol conquests drove a large number of glassmakers from Damascus and Aleppo to the West where they found employment in many of the glassmaking centres and had a great influence on design.

The rise of Venice

Venice, where glassmaking had been practised continuously since the end of the tenth century, became the greatest of these centres. A powerful guild was formed there which, in 1291, moved to the island of Murano because of the fire risk in Venice itself. Later the concentration on Murano enabled the industry to keep its closely guarded secrets from spreading elsewhere; the glassmakers were a privileged group, they were able to experiment with glass compositions and design, but attempted emigration by any guild member was punishable by death.

About this time additional substances were added to the batch in order to improve the transparency of the glass. Without additions, impurities, chiefly iron, in the raw materials cause an undesirable greenish-brown colour or lack of clarity. The Venetians used pyrolusite, the 'stuff that washes in the fire'; pyrolusite contains manganese which oxidizes the iron and improves the colour of the glass; it was also known as glassmaker's soap. By oxidizing the iron the manganese itself is reduced; the reduced form of manganese is colourless but when oxidized it is a strong purple. Manganese was used until quite recent times as a decolourizer and some old windows may be seen, particularly in Belgium and the Netherlands, where a purple colour has developed owing to long exposure to sunlight which has effectively oxidized the manganese back to the purple form.

By the early sixteenth century the Venetians using manganese as a decolourizer were able to make a glass as clear as fine rock crystal, known as Venetian 'cristallo'. At the same time they discovered how to make glasses in many beautiful colours—blue, white, green, purple and turquoise—and how to decorate with gilding and enamels, probably a contribution of the refugees from the Middle East. The lime content of the glass was reduced and the soda content was increased, thus increasing the range of temperature in which the glass could be worked. The workability of the molten glass was exploited to produce thin blown shapes in marvellous designs which quickly became fashionable throughout most parts of Europe (see plate 2). Again the production of thin blown glass probably received a stimulus from Syrian traditions of glass manufacture, brought to Venice by Syrian workers as a result of the Crusades.

Until the fifteenth century the Venetians had been in a favourable position for trade with both western Europe and the Byzantine Empire, but Constantinople was finally captured by the Turks in 1453 and the Venetian trade with the East was disrupted and began to dwindle. They therefore embarked on a campaign to increase the export of their products to the West. The *façon de Venise* or style in which the Venetians made their glasses, was immediately successful and spread rapidly to other European countries. It was developed by each country over a period of time to give distinct regional styles, a process which in the end led to the waning of Venetian influence. But for two hundred years, until the end of the seventeenth century, Venice dominated the world of glass.

The Venetians exported their glass to many countries but their knowledge was spread by workers who escaped from Murano during the sixteenth and seventeenth centuries, a period in which national states were emerging with controlled economies and with monopolies which made it possible for financiers to organize the Venetian methods of glassmaking on a large scale.

Although most industrial units of that period were very small by present-day standards, and remained so until the time of the Industrial Revolution, Venetian glassmaking seems to have been carried out on a larger scale, the more successful works probably employing as many as one hundred workers.

The English glass industry of the mid sixteenth century was somewhat backward, and it is of interest at this point to trace its development prior to the coming of the Venetians and to show how subsequently it gained an important position in European glassmaking.

Glassmaking in England

An old history of Bacton, Norfolk, provides evidence of glassmaking in England during the twelfth century. It mentions that King Stephen (1135-54) made a certain Henry Daniel, a 'Vitriarius' (*i.e.* a glassmaker), Prior of the influential monastery of St Benet's at Holme, Norfolk.

The chief centre of English glassmaking during the medieval period was Chiddingfold in Surrey. It is first mentioned in this connection in a grant of land to Laurence the Glassmaker, dated around 1226. The boundary of the area is defined in a deed of 1280 and it was thus possible to locate the site within which fragments of fused glass and broken crucibles have been discovered. Conclusive documentary evidence of a glass industry in the Weald is provided by the Exchequer rolls in 1351 and after this date it is possible to build up a fairly comprehensive picture from accounts, parish registers, deeds and parliamentary petitions.

The glassmakers moved frequently from place to place as they exhausted the available supplies of fuel. They preferred to burn beechwood and their glasshouses were also of wood, temporary structures with large openings left in the roof and walls for the escape of smoke and the free access of fresh air. Their furnaces are described in detail in Chapter 5 but it may be noted here that they melted in large open pots which were heated directly from below by a wood fire. Early tools were very simple (Figure 1), the most important being the blowpipe, the pontil rod, the shaping-tool and the bench or 'chair' on which the worker sat. These tools are still used for hand manufacture and were probably in use in similar forms from very early times. The blowpipe consisted of a wooden mouthpiece attached to about three feet of small-bore iron pipe. Nowadays the tubing for this iron pipe is drawn by machine, but until the early twentieth century blowpipes for glassware manufacture were hand-made by a smith from strips of wrought iron. The strips, which were about $\frac{1}{8}$th in. thick, 4 in. long and of a width to suit the diameter of the tube, were held at one edge and turned over to form a circular section. The longitudinal join of the tube was closed by hammering down the overlapping edge, the bore of the tube being preserved by a hardened former or 'drift'; the longitudinal joins were then fire-welded. The blowpipe was built up from these short sections of tube which were fire-welded to one another. The finished pipe was tested for leaks at the joins by pressing a heated end against a block of wood and holding the thumb over the other end. The wood-smoke was thus forced to find its way out through any faulty joins which could then be welded up. The glassmaker took up the molten glass on the end of the blowpipe by rotating it while dipping its end just below the surface of the glass, a process known as gathering. When a sufficient weight of glass had been gathered, it could be marvered and blown to form a hollow ball or bulb which was pressed into shape by the shaping-tool. This implement resembled

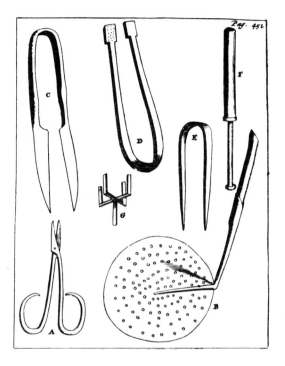

Fig. 1. The tools of the glassmaker, taken from *De Arte Vitraria,* Neri-Merrett, 1st Latin edition, Amsterdam, 1668. The simple tools used by the glassmaker have changed little over the centuries and implements like those shown in this seventeenth-century illustration are still in use today. Amongst the tools seen in the picture are the blowpipe for blowing the initial shape, pincers, tongs and shears for finishing the piece, and a holder for carrying away the glassware.

a large pair of sugar tongs, made of iron, with long blades in place of spoons. The pontil rod was an iron rod used for gathering small pieces of molten glass or seals and for holding the vessel temporarily during manufacture.

Small bowls, cups and bottles could be made with these tools and, later, bulbs of sufficient size were blown to be converted into flat window glass.

Briefly, the bulb was formed into an elongated cylinder, split from end to end when cold and opened out flat by reheating. A great deal of the business of the Wealden industry was concerned with making this brode (broad) glass but all coloured glass appears to have been imported from abroad which suggests that development of the domestic industry was not very advanced.

A new impetus came when in 1567 Jean Carré, a merchant from Antwerp but at that time resident in London, received a licence for twenty-one years from the Government of Queen Elizabeth I for making window glass in Britain. Many foreign glassmakers were willing to work in England and the successful exploitation of the monopolies was dependent upon them. The monopolies were conditional upon an undertaking that glass of all types made by the foreigners in England should be at least as good and cheap as the glass then being imported, and a promise that the English would be so taught that at the end of a stated period they should be able to make glass as well as the foreigners.

Carré established a group of glassmakers from Lorraine and Normandy in the Weald where they inevitably came into conflict with the existing population, especially the native iron workers who had a competing claim on the rapidly diminishing fuel supplies. The iron-workers had the added expense of charring the wood before they could use it and as long established residents they felt a grudge against the foreigners who could easily carry their craft from place to place in search of fresh fuel supplies. Moreover, in spite of their contract the newcomers refused to disclose the secrets of the craft. Their houses and work places were set on fire and they were even threatened with murder but they continued to prosper and worked out the period of their contract with Carré. In addition to broad glass it is probable that crown glass was made. Crown glass was flat glass made by widening the blown bulb, attaching a pontil rod to the side opposite the blowpipe, which was then broken off, and spinning the pontil rod so that the open-ended bulb expanded into a flat disc (see Chapter 6). Owing to their shape and the pontil mark the discs or 'crowns' could only be cut into small pieces, but they had a brilliant surface which had not come into contact with any other material during manufacture. This fire-polished glass was much more suitable for glazing than broad glass. The foreign glassmakers also undertook to make, and may have produced, some coloured glass.

When their contract expired, the Lorraine glassworkers left the Weald going west to Hampshire, Somerset, Gloucester and the Forest of Dean, then northwards up the Severn Valley into the counties of Shropshire and Staffordshire. The names of their descendants can be found as far afield as Newcastle and London, reflecting their far-reaching influence on the English glass industry. They were important, too, in the transition from wood to coal firing, of which more will be said later. Paul Tyzack, a 'gentleman glassmaker' whose family had come from Lorraine in response to the offer of a nine-year contract by Carré, was a key figure in this transition. He may have been the first glassmaker ever to use coal with complete success and he and his associates founded the famous Stourbridge glass industry.

Foreign influences did not only affect the making of window glass; a group of Venetian glassmakers led by Giacopo Verzelini was established in about 1575 in Broad Street, London, for the manufacture of drinking glasses. There had been contacts between Venice and England at least since 1399, when Richard II had allowed import of glass free of customs duty from two Venetian ships which had sailed into the port of London, an agreement which was renewed by Henry IV in 1400. In about 1550 eight Muranese glassworkers, who had come to England because of hardship in their own country, were forced by a group of Flemings and Englishmen to stay in England until they had fulfilled a contract with them and had paid alleged debts. The English were very eager to learn the secrets involved in the manufacture of fine Venetian crystal glass and to make it for themselves, rather than having to rely on the Italian workers or on expensive imports. However, a twenty-one-year licence was granted to Verzelini in 1575 for the making of drinking glasses 'suche as be accustomablie made in the towne of Morano'. Only a small number of glasses of this period remain which can be ascribed to Verzelini, but they show a very advanced technique and were made, as were Venetian glasses, of soda-lime-silica glass. Although the glasses are in the same style as Venetian glass of that period, they are much plainer and more functional than continental examples of *façon de Venise*, a tendency which was developed further at the end of the seventeenth century when lead crystal glass was introduced.

Coal as fuel for glass furnaces

The government was very pleased with the success of the industry, but the disappearance of valuable forests from which the glassmakers took their fuel supply caused increasing alarm. The wood was needed by the Navy for building ships, which were vital for the defence and prosperity of the country. The necessity for finding a new fuel, both for ironworks and glassworks, led to experiments in the heating of furnaces by coal in the early years of the seventeenth century and an agreement was made in 1610 between the Crown and a certain William Slingesby, by which the latter was given the sole right to use, in various industries, furnaces which would burn coal in place of wood, or, if already using coal, would permit greater efficiency in fuel consumption.

Slingesby intended that his furnace should be used for iron smelting and indeed Dud Dudley recorded in his *Mettallum Martis*, of 1665, that glass melting with pit-coal was 'first effected' near his house close to the Staffordshire and Worcestershire boundary, in an iron manufacturing district where coal had been mined since the thirteenth century. Developments in furnaces for iron smelting and glass melting were closely connected but

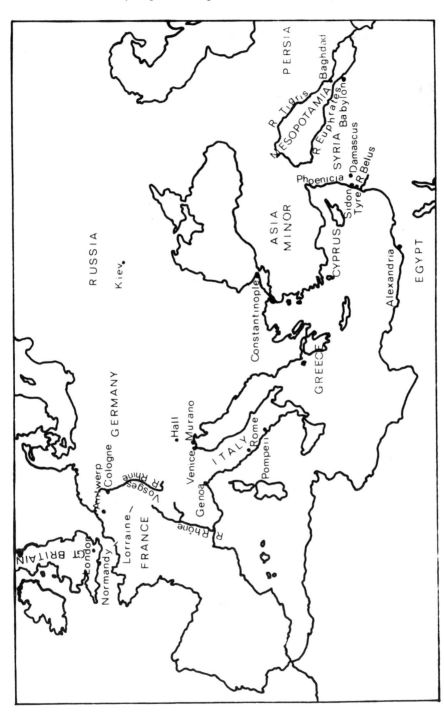

From its beginnings in western Asia and Egypt, glassmaking spread throughout the eastern Mediterranean area and into the Italian peninsula before the birth of Christ. During the time of the Roman Empire, glassmaking centres were established in Gaul, Spain, the Low Countries and Britain. British glassmaking, however, did not flourish until the sixteenth century with the arrival of foreign workers from Lorraine, Normandy and Venice. The map shows some of the important centres of glassmaking mentioned in the text, from the third millennium BC to the sixteenth century AD.

success was attained first in the glass industry. Before coal could be used for iron smelting it had first to be changed to coke to remove sulphur which otherwise weakened the metal. The first person to smelt iron ore successfully with coke was Abraham Derby in 1708, at Coalbrookdale, Shropshire.

In 1611 a Crown monopoly was granted to Sir Edward Zouch and a small group of gentlemen for making all kinds of glass using coal as fuel, and a decree was issued in 1615 in the name of James I forbidding the importation of glass, prohibiting the use of wood for the firing of glass furnaces and insisting on the use of coal. This decree resulted in the growth of glassmaking in coal-producing districts such as Stourbridge, where there was also an excellent supply of fire-clay for pot-making.

The group holding the Crown monopoly provided the necessary capital for experiments and manufacture and was allowed to make as much profit as desired. By 1618 one of the group, Sir Robert Mansell, had purchased the whole interest of the glass monopoly and for thirty-eight years he was the dominant figure in the development of the industry. He set up coal-fired glass furnaces in England and Scotland and caused the glassmakers to settle near the coalfields. The most important establishments were in the area around Tyneside, at Stourbridge and in southern Scotland. In London he kept the Broad street glassworks open for the production of glass in the Venetian style.

His other great contribution was the importation of barilla on a large scale from about 1621. Barilla was a Spanish plant ash containing a high proportion of soda which was exported in large quantities for use by glassmakers and soap manufacturers. Glass made from barilla was often judged to be of better quality than that made from other types of plant ash and the trade in barilla grew after the conclusion of hostilities with Spain at the beginning of the seventeenth century. The saving of timber following the universal adoption of coal for glass-melting and the establishment of window glass and bottle factories in many parts of the country were of great benefit to the community. On the other hand monopolies, many of which were established during the times of James I and Charles I, increased prices out of proportion to the profit gained by the Crown. Glass was not the only commodity to suffer; monopolies were granted on almost every article for domestic consumption including candles, dyes, fuel, hair powder, cutlery, wine, soap and salt. These abuses encouraged the glass-sellers to bind themselves into a trade protection society. The first Charter, dated 1635, was never implemented, probably because the process of incorporation was delayed from year to year owing to the political troubles which culminated in the Civil War. A Charter was finally granted in 1664, and the new Worshipful Company of Glass Sellers of the City of London was at once involved in efforts to improve the financial status of the livery and to suppress unregistered sellers.

The development of lead crystal glass

Complete monopolistic control of the glass industry came to an end in 1656 with the death of Mansell and many separate enterprises began to flourish. It was an excellent time for development as European tastes were turning from the Venetian glass, and in both Germany and England alternative styles were evolved which enabled these countries to assume dominant roles. A series of letters complete with diagrams written by a London glass-seller, John Greene, to his Venetian supplier (1667-72) illustrates the type of glasses that were replacing the elaborate Venetian products. Greene required Morelli to send glasses of 'sober forms and white sound mettall'. His drawings show large plain, straight-sided goblets with simple bases.

The heavy lead crystal glass perfected in England was admirably suited to these new styles. Lead glass is very transparent, the absorption due to the iron impurity being much less than in soda-lime glass on account of the oxidation of the iron. The refractive index is greater than that of Venetian crystal glass but it is chiefly the transparency which gives it its ability to sparkle in the light which falls upon its facetted or angular cut surfaces. Lead glass was quite unsuitable for making vessels in the elaborate fanciful styles of Venice but simple heavy styles decorated by cutting were encouraged by its optical properties.

Prior to the development of lead crystal glass, fine glassware was made in England which resembled in appearance the Venetian 'cristallo' and seventeenth-century English glassmakers made many experiments in order to perfect 'crystal glass'. Evidence in State Papers of 1666 shows that the Duke of Buckingham, a leading manufacturer and experimenter, was using saltpetre (potassium nitrate) as a flux in the manufacture of fine glass. The following extracts from Books of Rates (duties to be paid on imported goods) also show that potash was an alkali widely used at the time for the manufacture of glass:

1642:   'Ashes, vocat potashes: Barillia or Saphora, to make glasse'.
1657:   'Ashes called potashes: Barillia or Saphora to make glasse'.
1660:   'Ashes, vocat potashes: Barillia or Saphora to make glasse'.
1690:   'Pottashes: Barilla or Saphora'.

The use of potash rather than soda as the alkali in lead glass is preferable for blown ware and the perfection of lead glass suitable for making high-quality domestic and artistic glassware may have followed its adoption.

Lead had been used since ancient times in glasses. Several chapters of the important book by Neri, *L'Arte Vetraria,* published in 1612, were concerned with glass to be used for gem stones which was produced by the introduction of lead oxide, or 'calcined lead' as a flux to the sand. When it was realized that potash was preferable to soda in the production of a suitable lead glass for blown ware, difficulties still arose owing to the impurities present in the ash. The potash widely used at that time was a very crude material derived from the ashes of burnt plants and containing, amongst other things, a large percentage of lime. Pearl-ash, produced by purification of potassium compounds from the ashes of wood and land plants, may have been one of the alkalis used in fine quality English glassware. The ashes were mixed with water and boiled. This process of lixiviation was repeated many times; the liquid containing the soluble alkali was ladled out into shallow bowls, and after further settling the clear liquid remaining was evaporated until crystallization occurred. The crystals were then drained, dried and heated until hard. Finally they were crushed, sieved and added to the batch.

George Ravenscroft will always be remembered for his work in developing lead crystal glass during the last quarter of the seventeenth century. In his search for a rival to 'cristallo' he was no doubt strongly influenced by the English translation of Neri's book, published in 1662. Its translator, Christopher Merrett, a Founder Fellow of the Royal Society, had a wide knowledge of botany, chemistry and the history of technology and he was able to make valuable additions on the state of glassmaking. Merrett noted that 'glass of lead' was not made in England: 'tis a thing unpractised by our Furnaces ...' Ravenscroft, therefore, could have had very little practical information to draw upon. He had begun his experiments in London in the Savoy in 1673, employing Italian glassmakers and using imported quartz pebbles from the river Po. His glass may have been only a reproduction of 'cristallo' but the glass had a better appearance owing to the purer materials used. He obtained a patent for this glass and in the same year, 1674, the Worshipful Company of Glass Sellers of the City of London was sufficiently impressed by his work to set him up in a glasshouse at Henley-on-Thames, Oxfordshire, where he could carry on his experiments. They also undertook to find a market for his whole production, provided that vessels were made in specified shapes and sizes.

In 1676 Dr Robert Plot, a Fellow of the Royal Society and an Oxford don, wrote an account of a visit that he made to Ravenscroft's glasshouse at Henley-on-Thames:

> ... The invention of making glasses of stones or other materials at Henley-on-Thames was lately brought into England by Seignior da Costa a Montferratees and is carried on by one Mr Ravenscroft who has a patent for the sole making of them. ... The materials they used formerly were the blackest flints calcined and a white Christalline sand adding to each pound of these, as it was found by the solution of these by the ingenious Dr Ludwell Fellow of Wadham College, about two ounces of Nitre, Tartar and Borax. But the glasses made of these being subject to that unpardonable fault called crizelling caused by the two (sic) great quantities of the Salts in the mixture, which either by the adventitious Niter of the Air from without or warm liquors put in them would be either increased or dissolved; ... they have chosen rather since to make their glasses of a great sort of white pebbles which as I am informed they have from the river Po in Italy; to which adding

the aforementioned salts but abating in the proportions they now make a sort of pebble glass which are hard durable and whiter than any from Venice and will not Crizel. . . . And yet I guess that the difference in respect of Crizeling, between the present glass and the former lies not so much in the calx . . . but rather wholly in the abatement of the salts, for there are some of the first glasses strictly so called whereof I have one by me that has endured all trials as well as these last. . . .

It seems likely that at the time of Dr Plot's visit Ravenscroft was adding lead oxide to the batch in addition to the salts, but wishing to keep his new process a secret he attributed the improvement in the glass to the use of pebbles in place of flints. In the absence of lead oxide the large proportion of potash or nitre required to flux the silica would give a glass readily attacked by water and a network of surface cracks would result. Further experimental work on the proportions of the constituents appeared to be successful and in June 1676 the Glass Sellers Company announced that vessels made from the material would be satisfactory. A raven's head was adopted as a seal to distinguish the new glasses from the old and a three-year agreement was signed between Ravenscroft and the Company for a supply of the perfected glasses. Although no complete analysis seems to have been performed on any piece of certain Ravenscroft glass, tests on sealed examples and on pieces that are reasonably attributed to him give a lead content of the order of fifteen per cent.

By 1685 the production of the new glass was well established and members of the Glass Sellers Company were doing a thriving trade selling a wide range of fine quality glasses. The new lead glass was, and is still, known as flint glass, because flints were at first used in its manufacture, though fine sand was later substituted for the flints. It must not be confused with soda-lime glass of the same name, produced in England before this period, which also used flints instead of sand. Flint glass was one of the terms used as a trade-name by the Glass Sellers Company to describe Ravenscroft's newly developed lead glass, another being lead crystal. The term flint persisted, however, in connection with certain types of soda-lime glass, even though sand replaced flints, and it is still used today to describe white-flint bottles. The term crown, previously mentioned in connection with window glass, has also been used in varying contexts. In the field of optical glass, the two terms crown and flint were used to denote soda-lime glass and lead glass respectively, although the development of many new optical glasses has widened the usage considerably; for example, there are borosilicate crowns and barium flint glasses.

A great expansion of the British glass industry followed from the success of lead crystal glass and during the eighteenth century it achieved a leading position which it was to hold for a hundred years. From this period came the beautiful drinking glasses so prized by collectors (see plate 6 between pages 36 and 37). Glasses were made for all occasions: in celebration of military victories, to commemorate the Jacobite cause, for members of exclusive London clubs and as mementoes of christenings and weddings as well as for everyday use. The large number which survive show every variety of shape, size, decoration and type of stem: air twist, opaque, white or coloured twist. By the end of the seventeenth century the glass industry was flourishing as never before. Every country in Europe was making increasing amounts of glass, with the British industry gaining a leading role. Articles could even pass between countries before completion; wine glasses and goblets were sent from England to Holland for engraving prior to re-importation and sale.

The lead glass of this time did not have the hard crystal-like brilliance shown today; the articles had a faint greenish or blackish tone, more or less uniform throughout the glass, which is now regarded by collectors as a merit which adds interest to the vessel. Sometimes yellowish or amethystine tints can be seen; the tints may be attributed to variations in the iron or manganese contents and the conditions in the furnace. Early greenish and blackish tints were eliminated as time went on, and the brilliance of the glass was enhanced by suitable cutting, exploiting the relative ease with which lead glass can be cut and polished.

At the end of the seventeenth century, London was the leading centre with eleven glasshouses making lead glass. Typical of these was the Whitefriars Glasshouse, which dated from this time and which was situated on the grounds of the White, or Carmelite, Friars near the Temple. It remained there until the end of 1922, the only surviving works in the City, but it was then removed to Wealdstone, Harrow. For over two centuries, it is said, the furnace fires never went out, and when the works were moved to Wealdstone the new furnaces were lit by a flame brought from Whitefriars. An article in the *Whitehall Evening Post,* 1732, shows that the glassmakers were men of great spirit and independence:

> Yesterday a Press Gang went into the glass-house in White Fryars to press some of the men at work there, but they were no sooner got in but the (molten) metal was flung about 'em, and happy was he that could get out first, and in hurrying out they ran over their officer, who was almost scalded to death.

By the end of the seventeenth century lead glass was produced at centres throughout the country. Bristol and Stourbridge had four and five such houses respectively and Newcastle had an important one. The Tyneside area also turned out great quantities of utility ware in the cheaper soda-lime glass which was not necessarily inferior in appearance to that produced in lead crystal, and is often not recognized as soda-lime-silica glass.

English production was hindered only by a steady increase of taxation between 1745 and 1787 to pay for war against France. The tax was levied on glass by weight, and as the tendency had been to add more and more lead oxide, the production was checked. As a result many glassmakers moved to

Ireland where glass was free from duty and glassworks were set up in Dublin and Waterford.

Glassmaking in Germany and Bohemia

Lead glass was not the only new development of this period. Glassmaking in Germany and Bohemia had survived the collapse of the Roman Empire, the product being a coarse green glass, made with bracken ash, known as 'Waldglas' (forest glass), but in Bohemia from the fourteenth century onwards, attention was given to the preparation and selection of raw materials, especially quartz. Glass quality was improved by the addition of potash with a low iron content; manganese became an essential decolourizing agent.

Towards the end of the sixteenth century at the Prague court of Emperor Rudolf II, the techniques of engraving glass with copper or bronze wheels were developed. These techniques had been practised for many years both in Rome and in the German cities for the decoration of natural rock crystal, a crystalline form of silica, so the art was already highly valued. The Emperor gave the monopoly to engrave glass in this way to a diamond cutter, Caspar Leman. The art of engraving reached a peak during the seventeenth century but it was not until the end of this period that a glass was developed which deserved the elaborate decoration. Experiments had been in progress on the effects of adding lime to potash glass but some earlier products show a degree of degeneration which suggests that the process had not yet been fully mastered. By about 1680, a heavy, clear glass similar in appearance to natural rock crystal had been developed. It was a potash-lime glass which could be blown into thick heavy vessels; thus Bohemian crystal came into existence (see plate 7). The qualities of the raw material, the extreme skill and imagination of the cutters and the artistic climate of the period combined to produce outstanding works of art which were appreciated all over the world. The Bohemians were able to supply the demand through widespread sales organizations and were in fierce competition with the Venetians whom they eventually displaced all over Europe.

The German glassmakers had considerable success with coloured glass. The glasshouse of the Elector of Brandenburg in 1679 had as its director Johann Kunckel who, in 1679, had published the *Ars Vitraria Experimentalis,* a translation of the Neri-Merrett book with his own extensive additions. Although gold as a glass pigment was known prior to this date, no examples remain, and Kunckel was the first to develop and describe a reliable formula for the preparation of rich, ruby-red glass, using gold chloride as the colouring agent. Whilst this does not require as close a control of the degree of reduction as does copper ruby glass, it is necessary to reheat glass containing gold in order to develop a colloidal dispersion of gold particles upon which the colouring effect depends, and Kunckel must have made many experiments in order to perfect this glass. An eighteenth-century Potsdam formula probably describes his method. This prescribed taking a gold ducat beaten out thin and cut into pieces, and heating it with one and a half ounces of spirits of salt (hydrochloric acid), half an ounce of nitric acid and one dram of sal ammoniac until the gold was dissolved. The solution was then incorporated into the glass batch and after melting and shaping the glass was subsequently reheated to develop the ruby colour. The colour is much easier to control than the copper ruby and knowledge of the colour to be derived from gold chloride seems to have spread rapidly after its discovery. Others were experimenting along the same lines, for some time before 1679 Andreas Cassius of Leyden discovered that on adding tin chloride to gold chloride solution a purple powder, purple of Cassius, was precipitated. When added to the glass batch it produced a fine ruby colour if the glass was subsequently reheated.

Kunckel also developed an opaque white 'porcelain' glass, or 'milchglass', produced by adding burnt bone or horn, which contained phosphates, as the opacifier. Although such glass was made in Venice before 1500, it was the widespread admiration of porcelain throughout the eighteenth century that stimulated its production. Objects were made from porcelain glass in exact imitation of true porcelain and became immensely popular in Venice, Germany, Bohemia and England.

New demands on the glass industry

By the end of the eighteenth century the glasshouses which had flourished under the patronage of local rulers were no longer economic institutions. New factories had to be able to produce on a large scale wares that would appeal to a broad public. Large staffs of glassmakers were required and sometimes expensive equipment. For example, in the making of large sheets of glass by pouring and rolling as practised in France, many large flat tables, annealing ovens, overhead hoists, rolling apparatus and containers for transporting the molten glass were required for efficient operation.

The glassmakers of France had been organized for such large scale operations for over one hundred years. Firm official control had begun during the sixteenth century when Henry IV had given exclusive rights of ten to thirty years duration to Italians to make glass within four main geographical areas; Paris and its surroundings, Rouen with Normandy, Orleans and the Loire country, and Nevers with the Nivernais. From 1665 onwards, Louis XIV and his minister Colbert, by vast expenditure and the use of absolute royal powers, concentrated all glassmaking resources into creating a flat glass industry. The need for flat glass became acute when Colbert had to face the enormous cost of installing the magnificent Venetian mirrors in the Galerie des Glaces at Versailles (see plate 8). He attempted to import Venetian glassmakers into France, but his efforts were unsuccessful, two of his most skilled Venetian workmen mysteriously dying within a few weeks of each other.

However, in 1688, Bernard Perrot, holder of the royal monopoly for glassmaking in Orleans, produced flat glass by casting, which gave larger sheets with surfaces more suitable for polishing than those produced by any blown glass method. Very few details of this invention are available but it probably included a method of rolling to produce a flat surface with subsequent annealing, grinding, and hand polishing. The famous St Gobain factory became the *Manufacture Royale des Glaces de France* in 1693, and soon the French replaced the Venetians as the most important exporters of fine mirror glass.

The French lagged behind the English, Germans and Italians in the making of decorative glassware but the Baccarat factory was established in 1765, and other factories followed towards the end of the century. The French concentration on flat glass manufacture began in the Middle Ages when the finest crown glass in Europe was made in Normandy and the best broad glass in Lorraine. France at that time produced wonderful coloured glass for windows, which was widely exported.

By the end of the eighteenth century expanding economies increased the demand for glass. The establishment of large, well-equipped factories was accompanied by the development of well-organized sales and distribution. At the same time the Industrial Revolution was taking place and the science of chemistry was beginning to develop. At the beginning of the period alchemy still lingered, and such chemical theory as existed, was based more on speculation than observation. Then, within a relatively short period, many of the ideas which form the basis of modern chemistry were developed. The chemical industry which, from its earliest years, was concerned on a large scale with the production of alkali from brine, began to grow rapidly. The major customers of the alkali industry were the glass and soap manufacturers. Just as the glassmakers had found a fuel other than wood, so they now had available a replacement for natural alkali and thus the beginning of the nineteenth century may be viewed as the start of a new era for the glass industry.

Bibliography

1. *A History of Technology*, Vol. IV, (eds) C. Singer, E. J. Holmyard, M. R. Hall and T. I. Williams, Oxford University Press, 1958.
2. *Glass*, G. Savage, Weidenfeld and Nicolson, 1965.
3. *From Broad-glass to Cut Crystal. A History of the Stourbridge Glass Industry*, D. R. Guttery, Leonard Hill, 1956.
4. *The Glass Industry of the Weald*, G. H. Kenyon, Leicester University Press, 1967.
5. *Glass*, E. Dillon, Methuen and Co., 1907.
6. *Glass throughout the ages*, R. J. Forbes, *Philips tech. Rev.*, 1960/1, **22** (9/10), 282.
7. *Masterpieces of Glass*, The British Museum, 1969.
8. *Bohemian Engraved Glass*, Z. Pešatová, Paul Hamlyn, 1968.

9. *Economic and Social Aspects of European Glassmaking Before 1800,* A. Polak, review paper, VIIIth International Congress on Glass, London, 1968.
10. *Glass Through the Ages,* E. B. Haynes, Pelican Books, 1948.
11. *Glass-making in England,* H. J. Powell, Oxford University Press, 1923.
12. *The early history of glass-making in the Stourbridge district,* D. N. Sandilands, *J. Soc. Glass Technol.,* 1931, **15,** 219.
13. *Stourbridge cameo glass,* J. Northwood, *J. Soc. Glass Technol.,* 1949, **33,** 106.
14. *A History of English and Irish Glass,* W. A. Thorpe, Medici Society, 1929.
15. *The Worshipful Company of Glass Sellers of London,* G. A. Bone, 1966.
16. *A notable seventeenth-century contribution to the literature of glassmaking,* W. E. S. Turner, *Glass Technol.,* 1962, **3,** 201.

2

The glass industry in the nineteenth and early twentieth centuries

During the nineteenth century and the early years of the twentieth century, glassmaking methods changed from those which had been practised with little alteration since the beginnings of the craft thousands of years previously, to those which form the basis of the mechanized processes of today. The introduction of radically different methods of glass-melting, the development of automatic processes of forming glass, the changing economic conditions and the increasing demands of customers for more varied types of glass, all contributed to these changes.

The developments in different countries occurred in response to different circumstances. In continental Europe the early recognition of the contribution that science could make to the improvement of industrial processes led to a rapidly growing chemical industry and to the encouragement of scientific research in universities and the Technischen Hochschulen. In Britain, the repeal of onerous taxes and excise duties was followed by a great increase in the demand for flat glass; and in the United States, shortage of skilled labour and the activities of the Trade Unions encouraged the mechanization of many processes.

The conditions of glassmaking in Europe during the first part of the period were very different from those in Great Britain. Britain, the pioneer in the use of coal as a fuel, had used and developed new methods of melting since the beginning of the seventeenth century. Two hundred years later some of the glass industries of Europe were beginning to attempt to melt with coal, but the vast majority of glass-workers still burned wood.

The continent of Europe

Raw materials

In Europe at the beginning of the nineteenth century, the glassworks were generally 'forest' industries, very often at a primitive stage of development technically, if not artistically. The flux was mainly potash produced by burning local wood, lixiviating the ashes and evaporating the solution; this residue was mixed with silica-sand found locally. Potash remained a major source of alkali even after coal had replaced wood as fuel for, by that time, it had become available from the rich Stassfurt potash deposits which were developed in 1861. These deposits, which have a total estimated thickness of

2500 feet, eventually made the Western European glass industry independent of alkali from other sources.

Glassworks near the sea, such as those in Normandy, used sea plants as sources of soda. The Russians used impure soda made from natural sodium sulphate by reducing it with coal. The sulphate, known as 'goudjir', had been discovered in large quantities during the eighteenth century in the salt lakes of Siberia. When sodium sulphate was produced artificially during the nineteenth century, a similar addition of a reducing agent, such as charcoal or powdered coal, was found to be necessary; the reaction between sodium sulphate and silica-sand proceeds very slowly below about $1350°C$ but is accelerated by reducing the sulphate to sulphite or sulphide. Any unreduced sulphate is liable to remain on the surface of the glass as gall but if too much carbon is added a foam is formed and the glass may acquire an amber tint.

In France there had been a general change from potash to soda at the end of the eighteenth century, for alkali from marine plants, mainly Spanish barilla, was found to give better quality glass although it still had a greenish or bluish tint. Barilla had been used in Britain at least since the early seventeenth century when Sir Robert Mansell imported it on a large scale for glassmaking. However, by 1775 the prices of potash and Spanish barilla had risen so much that the French Academy offered a prize of 12,000 francs to the author of the best process for making soda from common salt. In 1787 Nicolas Leblanc succeeded in making unrefined soda by heating sea salt with sulphuric acid to give sodium sulphate or salt-cake, and hydrochloric acid. The salt-cake was then heated with chalk and charcoal to produce sodium sulphide and sodium carbonate. This 'black ash' was leached with water and sodium carbonate was recovered from the solution by evaporation. Sodium carbonate produced by the Leblanc method was widely used in glassmaking for over fifty years. Figure 2 shows the manufacture of soda by the Leblanc process.

Leblanc's own factory was closed during the French Revolution but the method spread to other countries. In 1823, James Muspratt built a plant in Liverpool followed by one at St Helens for the production of soda by the black ash process, and manufactured soda very quickly replaced imported barilla and ash from kelp or seaweed as the major source of alkali for British glassmaking. The importation of barilla continued until the early nineteenth century, but from about 1730 to 1830 large amounts of alkali were also produced by burning kelp. At the end of the eighteenth century 'kelping' was an important industry in Scotland employing thousands of people; Lord Macdonald of the Isles made £10,000 in a year out of his kelp shares. The price of barilla was continually increasing and kelp ash became a valuable source of alkali for the British glass industry, only declining in importance as manufactured soda became widely available.

Other industries such as the mining and metallurgical industries competed for the dwindling timber supplies which were the sources of fuel and plant ash and the soap makers and bleachers multiplied their demands for alkali during the great industrial expansion at the end of the eighteenth century.

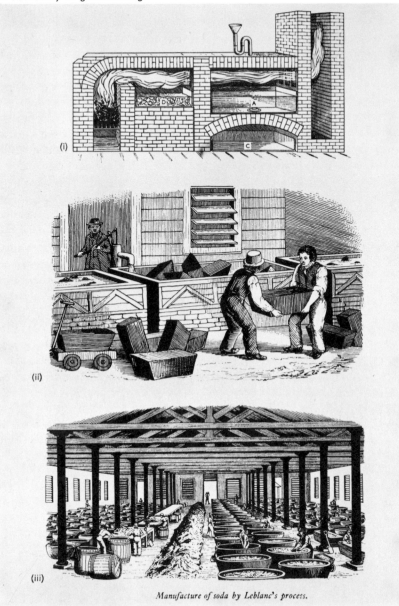

Manufacture of soda by Leblanc's process.

Fig. 2. Manufacture of soda by Leblanc's process. Shown in (i) is a section of a salt-decomposing furnace. Common salt and sulphuric acid are brought together in the lead-lined compartment A and are heated for several hours to produce a mass of sodium sulphate and fumes of hydrochloric acid. The mass is then pushed into C and after being thoroughly mixed with chalk and charcoal the reaction is completed in D, the hottest part of the furnace. In (ii) the resulting 'black ash' is leached with water in vats and (iii) the sodium carbonate is crystallized out from the solution in large tubs. The crystallizing house shown in the diagram is that of Chance Brothers at Oldbury, Worcestershire, where soda was prepared for use in the production of plate glass.

During the 1830s and 1840s, salt-cake from the first stage of the Leblanc process was increasingly used by glassmakers in place of the final more expensive product, sodium carbonate, but sodium sulphate demanded higher melting temperatures and increased the need for improved furnaces.

In Belgium in 1863, Solvay introduced the manufacture of sodium carbonate by the ammonia-soda process. The resulting fall in price led to soda becoming once more the major source of alkali in glassmaking. Ammonia and carbon dioxide obtained by heating calcium carbonate were allowed to react with salt solution causing precipitation of soluble sodium bicarbonate; this was then heated to give sodium carbonate. The ammonia could also be recovered from solution by the addition of the lime which remained when the calcium carbonate was heated. Solvay soda was often mixed with sodium sulphate by the glassmakers and some salt-cake is still added to many batches today because it helps to prevent the formation of siliceous scum on the surface of the molten glass; if soda is used alone it reacts very rapidly with the rest of the batch materials and the resultant sudden evolution of gas bubbles sweeps unreacted silica to the surface where in the absence of sufficient alkali it forms a scum. Adding salt-cake also helps to remove small bubbles from the molten glass, a step known as refining.

Improvements also occurred in other batch materials. The early glassmakers used a surprising variety of rocks and sands for their silica supply. The Russians used red and white sands and white sandstone; the glassmakers of Bohemia in the Austro-Hungarian Empire melted basalt, granite, felspar and obsidian. Sometimes their melts were satisfactory but generally the complexity of the material, the many impurities it contained and the variation in composition, produced very poor glass. All felspars are complex materials consisting mainly of alumino-silicates of potassium and sodium; granite is a mixture of quartz, felspar and mica, basalt is a mineral of volcanic origin chemically resembling felspar and obsidian is a lava or volcanic rock containing from sixty-five to seventy-five per cent silica. Obsidian was also used in France during the late eighteenth century as a constituent of bottle glass but it was later abandoned owing to lack of homogeneity.

In Germany, glassmaking was improved by increasingly careful selection of sand. Good glassmaking, and relatively iron-free, sand was mined, for example, in Silesia and near Aix-la-Chapelle. The iron concentration was not so important for green bottle glass, and granite and trachyte, with their high alkali contents, could be used as cheap raw materials. Limestone of sufficient purity and supplies of red lead for crystal glass were also available; with good batch materials and coal available the German glass industry grew rapidly.

The importance of lime as a stabilizing ingredient was recognized in the latter half of the eighteenth century when P. D. Deslandes of the St Gobain glass factory restored durability to glasses melted from purified ash by adding up to six per cent of lime. His experiments were not widely publicized and it was not until pure manufactured alkali became generally available in the early nineteenth century that lime became widely recognized as an essential ingredient of the batch. Until this time the vegetable ashes which were the source of alkali also contained large proportions of stabilizing materials such as lime, alumina and magnesia, and it was considered bad practice to add lime to the batch because the resultant glasses, which contained excess lime, often tended to devitrify. A handbook of 1835 notes that 'lime has only in the most recent years come to be recognized as an important constituent of the batches for most kinds of glass'. In 1830 Dumas found that a glass became more resistant to moisture attack as the composition approached the proportions of one molecule of alkali and one molecule of lime to six molecules of silica.

Fuels and furnaces

Wood was the only fuel used until the appetites of the furnaces of the glassmakers and the iron-workers in Great Britain became so enormous during the early seventeenth century that they caused government intervention which, in turn, encouraged investigations into the use of coal as a fuel. On the Continent the woodlands were much more extensive than in Great Britain and coal-firing was not widely adopted until the late nineteenth century. In Chapter 5 the story of the development of glassmaking furnaces is told in some detail; the early writers gave drawings of furnaces which hardly allow the exact method of operation to be deduced but Figure 3 shows one of the many furnaces which was widely used on the Continent, the 'German' furnace. First described in 1556 by the German writer Georgius Agricola, it remained in use throughout the eighteenth century and probably into the nineteenth century. The eighteenth-century picture shows a furnace with a bottom compartment in which the fuel was burnt, a siege on which melting pots stood and an upper floor for annealing.

The coal-fired English cone furnace provided an excellent draught and furnaces similar to that shown in the early nineteenth-century engraving reproduced in Figure 4 were still in use in the first half of the century. But from 1856 onwards the Siemens regenerative furnace was being developed and it is still the mainstay of the glass industry. Its introduction, the development of new refractories and the growth of the chemical industry enabled constant supplies of good glass to be available to feed the new glass-forming machines.

Artistic glassware

Just as impressive as the developments in raw materials and furnaces were those in artistic glassware. In continental Europe, Britain and the USA, artistic glassware was produced on a large scale, often involving complex technical processes.

Fig. 3. The 'German' furnace, 1752. This type of furnace was in widespread use on the Continent of Europe from the sixteenth to the nineteenth century. The fire was in the lower compartment, the glass pots were worked through holes in the walls of the middle compartment and were heated by flames rising through the hole in its floor, and the finished articles were slowly cooled in the upper compartment.

View of Aston Flint Glass Works belonging to Messrs W. Gillons, Birmingham.

Fig. 4. The English cone furnace was a familiar feature of the eighteenth and nineteenth centuries. The tall cone ensured ample draught to the coal-fired furnace resulting in higher melting temperatures than were possible with wood-firing. The transport of raw materials, coal and finished articles was facilitated by the widespread network of canals, one of which is shown in the foreground of the picture, and later by the coming of the railways.

The Bohemians maintained a continuous tradition of craftsmanship in artistic glass and, during the nineteenth century, they began to experiment with glasses of different colours and textures. One of the major aims, as in other countries, was to make glass which resembled natural stone. For example, between 1840 and 1850 a glassmaker, Wilhelm Kralik, created a coloured alabaster glass, turquoise alabaster glass and a green and pink chrysoprase alabaster. A fluoride opal glass was made in 1860 using Bohemian cryolite but by 1880 the sources of this mineral were in the hands of a monopoly and prices rose steeply. Because of these difficulties new compositions for opal glass were developed containing phosphates or tin oxide. The Bohemians were very good at imitating gemstones and they made textured glass surfaces such as the scaly hornglass. New finishing processes were also developed. Frederick Egermann was the first to polish opal glasses, improving the appearance and leading to a considerable increase in sales. His son Ambrose introduced cork polishing wheels, and electrically driven polishing apparatus was perfected by the Bohemian, Kreybich, at the Reich works, Krasna, where it was introduced commercially in 1887.

Ludwig Lobmeyr of Vienna had a great influence on Bohemian production of artistic glassware during the second half of the nineteenth century. He was a pioneer of reform in both artistic and industrial glass. He revived the seventeenth-century skill in design and decoration which had greatly declined by the beginning of the nineteenth century because the Bohemians had been attempting, with glass of very different composition, to imitate the rich style of cutting practised by the English on lead glass. Lobmeyr also developed or rediscovered many kinds of coloured glass and revived the technique of producing a black enamelled design between two layers of glass, a technique widely used in previous centuries in Bohemia and Germany.

The surface treatment of glass was explored in France by Emile Gallé, a man with great artistic talents who reached the height of his public reputation at the Paris Exhibition of 1889. Plate 9 shows an example of his work. He made many experiments with various colouring agents, paying particular attention to chemical composition, the duration of heat treatment, and the chemical nature of the furnace atmosphere in which his glasses were melted or heat-treated. Similar experiments were made in the USA by Louis Comfort Tiffany who was famous for his glass in the 'art nouveau' style.

In Great Britain the craftsmen of the Stourbridge district produced a wide variety of artistic and utilitarian glassware which often showed considerable technical expertise. Plate 10 shows one of their highly distinctive products. Their most productive period was from the 1870s to the outbreak of the First World War. It was in Stourbridge that John Northwood worked upon his copy of the Portland Vase and revived the art of cameo glass. The glass artist, Frederick Carder, also worked as the chief designer for Stevens and Williams of Brierey Hill for more than twenty years before he went to the United States and founded the Steuben Glass Company of New York. The liveliness of the industry is typified by the pattern books of Thomas Webb and Sons of Stourbridge which contained some 25,000 items in their main series between 1837 and 1900, some of which are still in production today.

Great Britain

At the beginning of the nineteenth century Britain possessed a well-developed and organized glass industry. Coal firing of glass furnaces had been practised since the early seventeenth century and the very efficient cone furnace had been developed in which higher temperatures could be attained than in the older wood-fired furnaces in general use in other countries at this time. Coal firing seems to have spread very slowly from England, giving the English glassmakers a distinct advantage over their rivals. For example, wood was still in use in 1829 in the French glassworks at St Gobain, a well-organized works with advanced manufacturing methods, although experiments with coal had been in progress since the eighteenth century. Georges Bontemps, in his book of 1868, *Guide du Verrier,* said:

> It is generally known that the English glasshouses are huge cones which surround the furnace. The English coal furnaces are thus situated under big chimneys which encourage energetic combustion in a way that is impossible in our French furnaces, which have a louvered opening above the furnaces for the combustion products.

The Industrial Revolution started in England during the latter part of the eighteenth century, but this did not radically affect the glass industry in its early stages because motive power was not generally required in the glassworks of that time. The impact of mechanization is shown best by its development in the American glass industry under pressure of the demands of organized labour, but other changing social and economic conditions of the first half of the nineteenth century had marked effects on British industry.

Window tax and excise duties on glass

A tax was placed on windows during the 1690s in order to supply money for the wars of William III. Although factories, warehouses and uninhabited houses were exempt, windows were regarded as such luxuries that the occupier of a house with ten windows had to pay an annual tax of 8s 4d (about 42p) in 1776, rising to £2 16s 0d (£2.80) in 1808. This high rate continued until 1825 when the tax was halved, in 1825 houses with seven or fewer windows were exempt. The reduction coincided with the building boom of the 1820s and stimulated demand for glass. But according to manufacturers, the tax still limited demand and they said that in both northern and southern Europe houses had more than twice as many windows as in Britain.

An excise duty which had far-reaching effects on the manufacture of glass had also been imposed during the 1690s. Just as monopolies had been a profitable source of revenue during the sixteenth and seventeenth centuries, so during the eighteenth and nineteenth centuries the government regarded the glass industry as an inexhaustible fund to draw on in times of war and shortage. In spite of this handicap the industry survived and prospered by the continuous introduction of new techniques and by successfully forecasting and supplying the demands of a rapidly growing urban population.

A glass duty was first imposed by statute in 1695 and made perpetual the following year, but it was so high as to discourage manufacture and was soon reduced by half. The situation continued to deteriorate and by 1698 the consumption of coal had been so reduced, and unemployment had increased so much, that the duties were repealed. In 1746 duties were again levied but they were also imposed on imported glassware, the government presumably hoping to reduce the effect of the duties on the glassmakers. The Act of 1746 required a record to be kept of all furnaces, pots, pot chambers and warehouses, and due notice to be given when pots were to be changed. This Act was the first of many which covered all aspects of production and marketing. Penalties were severe, from £20 to £200, and in 1826 another burden was added in the form of an annual licence of £20 for every glasshouse. In the same year the regulations were applied for the first time to Ireland as a result of which many of the flourishing glassworks established there to avoid the excise duties began to decline.

The excise regulations of the reign of William IV give a vivid impression of the difficult conditions in the industry:

> (The glassmaker was) to mark and number every workhouse, pot chamber, pot hole, lear, warehouse-room and other place so entered and made use of; and not to use any pot for preparing or making glass without first giving notice thereof to the proper officer. . . . Every annealing oven, arch or lear . . . for annealing flint glass to be made of rectangular form with the sides and ends perpendicular and parallel to each other and the bottom thereof level and with only one mouth or entrance, with a sufficient iron grating affixed thereto and proper locks and keys and other necessary fastenings. Twelve hours' notice to be given before beginning to fill or charge any pot for making glass . . . the hour at which it is intended to begin, the weight of materials to be used and species of glass to be made. Six hours' notice to be given in writing of intention to heat any oven, arch or lear into which any glass is intended to be put for annealing. In all glass chargeable by weight, the grating to the annealing arches to be closed and securely locked and sealed by the officers immediately after all the glass or ware have been deposited therein. In weighing glass the turn of the scale to be given in favour of the crown.

The excise officers who supervised the restrictions were quartered in the glassworks where they caused great annoyance to the manufacturers, one of whom said:

> Our business premises are placed under the arbitrary control of a class of man to whose will and caprice it is most irksome to have to submit. We cannot enter parts of our own premises without their permission.

Even glassworks which were not in operation were inspected at least once a day. The cost in terms of government officials and wasted time in the glassworks must have gone a long way towards outweighing any benefits that the government might have gained from the duties. The import duties did, however, in part protect the window glass industry from foreign competition and window glass manufacturers were generally in favour of their retention in a modified form. Lucas Chance of Chance Brothers, Smethwick, expressed the opinion that increased consumption of window glass would be encouraged to a far greater extent by the abolition of the window tax in preference to the excise duties.

Effect of duties on technological development

The duties seriously delayed technological innovation; for example, the optician John Dolland said that it was impossible under the regulations of the Excise to make experiments to produce the glasses of various refractive indices that he specially required. He had obtained permission from the Treasury to continue his promising experiments at the Spon Lane works of Chance Brothers and to have crown glass made there of sufficient thickness to cut up for lenses. However, the crown glass was made by the glassworks' employees and came under the charge of the Excise supervisor. An Excise law forbade the production of crown glass exceeding one-ninth of an inch in thickness, which was too thin for Dolland's purposes. The prohibition had arisen originally because the manufacturers had exported thick crown glass and had called it plate glass which carried an advantageous refund or 'drawback' on the export dues, while crown glass did not. Again, Cookson and Co. of Newcastle and South Shields were asked by the Northern Lighthouse Board to try to construct a Fresnel polyzonal, or stepped-surface, lens. They succeeded using plate glass but their plate had to be five-eighths of an inch thicker than that recognized as the limiting thickness for plate glass by the Excise laws. The duty on such thick glass was prohibitive, £1 18s 0d (£1.90) per hundredweight greater than that for plate glass. Cookson's subsequent petition to Parliament for reduction of duty probably did not get far as such a reduction would have required an Act of Parliament.

In the bottle trade the experiments of Frederick Fincham, who had made acid-resistant bottles and phials for chemists from his green bottle glass, were abruptly stopped by the Excise Commissioners. They said:

He was informed that his work could not be allowed to continue because he produced an article so good that it could not be sufficiently distinguished from flint glass, the danger being that this article, which for a great variety of purposes was admitted to be in all respects as good as the comparatively highly taxed and therefore high-priced article of flint glass, would be substituted for that description of glass to the detriment of the revenue however much the substitution might conduce to the convenience of the public.

Fincham's phials were also illegal because no bottle-maker might make anything with a capacity of less than 6 oz.

Compositions of bottle glasses were also restricted by law to certain raw materials and with 125 per cent total duty on the manufacturing process a bottle-maker was not inclined to make experiments, especially with coloured glasses.

The number of different duties was matched by the many ingenious ways of avoiding payment, and corruption frequently occurred. A Lancashire crown glass manufacturer had one side of his annealing oven made removable and articles could be taken out with long forks operated by a lever. Officers were given no bounty if they caught an offender and William West of Mackay West, Lancashire, used to bribe the excise officer on duty. It was estimated that 600 pieces of glass were fraudulently removed from the annealing oven each week and as the duty was imposed according to the weight of glass this represented a considerable sum of money. West was eventually exposed and had to pay £5000 to the Commissioners; his firm became bankrupt.

Excise duties and window glass manufacture

The excise duties were probably directly responsible for Britain clinging to crown glass manufacture long after the establishing of the superior method in which a cylinder of glass was blown, split and flattened in a special oven. Duty on window glass was levied by weight, but the glass was sold by size and quality. A thin disc was easier to make than a thin cylinder and the advantage remained even after allowance had been made for the greater wastage in crown glass manufacture. The thin discs bore less duty but the size and quality were about the same; even so the duty was more than half the selling price. The introduction of cylinder or sheet manufacture was also delayed both by the preference of the English for the bright, untouched fire-finished crown glass surface and by the initial capital expenditure necessary for the new process.

Although the excise duty favoured crown glass for so long, paradoxically it eventually encouraged the introduction of sheet glass. In 1813 the only type of window glass exported was crown glass, on which a manufacturing duty of £3 13s 6d (£3.67½) per hundredweight was charged. This duty was levied on the whole disc, but because the disc was then cut up into panes, the

Fig. 5. The Crystal Palace was built to house the Great Exhibition of 1851 and to display for the first time 'the Works of Industry of all Nations'. The building itself, designed by Paxton, contained nearly one million square feet of glass supplied by Chance Brothers of Birmingham.

manufacturer was given a rebate of £4 18s 0d (£4.90) per hundredweight of cut glass panes for export to compensate for the manufacturing duty and for wasteage in the cutting. Cylinder glass, which was of little importance at the time, was allowed the same rebate although there was not so much wastage in cutting and thus part of the rebate became a bounty to exporters of cylinder glass. Chance Brothers realized this and in 1831 started to make cylinder sheet glass chiefly for export. In sixteen months they had made a profit on the excise duties alone of about £1270 which they used to offset part of their development costs. By the time that the Excise Commissioners had realized what was happening and had acted to reduce the rebate, the process was well established and Chance Brothers were ready to supply glass for the great building booms of the 1830s and 1840s.

Repeal of excise duties—1845

After ten years of strong and continuous protests from the manufacturers the duties were repealed in April 1845. The industry immediately entered a period of rapid growth and, to increase the labour force, French and Belgian workers were brought in. The foreign workers were able to dictate wages and terms of employment and the employers found that they had 'exchanged the Excise for a much severer taskmaster (namely), our own men'. Between 1844 and 1846 wages had increased by thirty per cent and were continuing to rise. The very high wages paid to the foreign workers were not at first paid to the British employees, and the British soon demanded better pay and conditions. They were generally bound on contract for seven years at a time, while the foreign workers only worked for their masters for one campaign, or for the life of the furnace. This led to much trouble with contract breaking and firms were constantly involved in legal wrangles over employees.

In 1851 the Crystal Palace was built and nearly one million square feet of glass were supplied by Chance Brothers (see Figure 5). There had been much competition between Chance Brothers and Messrs Hartley of Sunderland for the contract to supply this glass, both firms having recently developed the capacity to make the large quantities required. Chances had obtained the services of the eminent French glass technologist Georges Bontemps in 1848 and he had brought many foreign workers skilled in the production of cylinder glass to England. Hartleys had developed their own process for making sheets of thin cast plate-glass in 1847, but the Exhibition jury decided that its strength had not been proven in use and Chance Brothers were given the contract.

After the Exhibition new uses for glass were increasingly found in public and domestic buildings. Coloured, stained and painted glass became very popular and the work of glaziers and glass painters was prominent in the great international exhibitions which were a feature of the late nineteenth century.

The North American glass industry

In the United States major developments took place in semi-automatic and automatic processes during the nineteenth and early twentieth centuries, but at the beginning of the nineteenth century the industry was in a primitive state. The history of glassmaking during the Colonial period was one of continual failure; the early settlers set up glassworks but eventually they were all abandoned. There was little domestic demand for glass, the simple houses had no glass windows and most domestic utensils were made from cheaper materials such as wood. There were adequate supplies of raw materials but a great shortage of skilled workers, and little inducement for English craftsmen to settle in America because the conditions at home were much more advantageous. The colonists who could afford the luxury of fine glassware bought English glass in preference to the native products.

At the beginning of the nineteenth century there were probably not more than twelve glasshouses in North America. Foreign supplies were cut off during the American War of Independence (1775-83), and this cut in supplies produced a temporary increase in the demand which declined with the return of peace. Nevertheless, from 1786 to 1800 glassworks were established in at least six states and these formed a foundation for the development of the American glass industry. The success of these firms was in part due to government help in the form of bounties, protective tariffs and relief from military duties. Imports were restricted during the years 1808-14 when the wars in Europe disrupted supplies. The war of 1812 with Canada cut the supply of glass from England and the American industry expanded. But peace brought a flood of English imports aided by English export subsidies and in the next few years the American glass industry was suffering a general depression.

By 1820 the industry was beginning to expand again and this expansion was destined to be far more permanent than that of 1812. It was influenced greatly in its early stages by the government's protective tariffs; the domestic production of window glass soon dominated the market and by 1831 imports were almost eliminated.

The techniques of 1820 were basically those which had been used since the foundation of glassmaking in the country. Many of the early glassmakers were German immigrants and the designs of the furnaces were similar to those of Germany, burning wood which was cheap and plentiful. Coal was available in certain places but it was scarce in the areas where glass manufacture was first established and although English immigrants tried to introduce the English coal-burning cone glass furnace for flint glass manufacture, they eventually adopted the continental types of furnace.

The best raw materials had not been discovered by 1820 and many works were engaged in costly and sometimes ruinous experiments. Glass production west of the Alleghenies was started at Pittsburgh in 1797, an area which was

Plate 2. A Venetian style tazza, sixteenth century AD, southern Netherlands: height, 15.9 centimetres; diameter, 15.9 centimetres. During the sixteenth century the Venetian influence spread throughout Europe and much of the glass of this period is delicate and ornate, reflecting the complex forms produced by the Venetian craftsmen.

Plate 1. Oinochoe, fifth century BC, from Camiros, Rhodes: height, 10.5 centimetres; diameter, 6.8 centimetres. The oinochoe was a popular form of container in the eastern Mediterranean area prior to the Christian era and was made by the laborious process of covering a shaped core with glass, either by dipping the core into the molten glass or by winding threads of glass around it. The oinochoe was decorated by applying coloured threads of glass to its surface and pressing them into the surface by rolling. After the addition of the handle and foot the core was chipped out.

Plate 3. Roman blown glass handled flagon, late second or early third century AD from Bayford, Kent: height (to rim), 20-20.8 centimetres; diameter (greatest), 15 centimetres. The invention of glass blowing made possible the production of thin transparent vessels on a much larger scale than had been possible by the older forming methods. This elegant flagon shows the degree of skill reached by the Roman glassmakers in the making and forming of glass.

Plate 4. The Portland Vase, late first century BC or first century AD: height 24.5-24.8 centimetres; diameter (greatest), 17.7 centimetres. The Vase consists of a dark blue inner layer and an outer opaque white casing of glass into which the design has been cut. It combined the recently developed skill of glass blowing with the traditional Egyptian art of carving in relief on stone.

Plate 5. Syrian mosque lamp, 1330-45 AD: height, 33 centimetres. Many perfect examples still survive, having hung for centuries in the mosques for which they were designed. They embody the Islamic skill in decoration by enamelling and gilding.

Plate 6. Eighteenth-century English wine glasses of lead crystal. During this century the English glassmakers were famous for these beautiful glasses with their delicate twisted stems and fine engraving. The lead glass from which they were made was very transparent giving them the ability to sparkle in the light like diamonds.

Plate 7. Bohemian crystal glass baluster goblet, early eighteenth century. The Bohemians developed a heavy, clear glass similar in appearance to natural rock crystal which could be blown into thick, heavy vessels and engraved with traditional designs.

Plate 8. The Galerie des Glaces, Versailles. This great hall in the Palace of Versailles, built for Louis XIV in 1678-84, has a complete wall of mirrors which can be seen in the left hand part of the picture.

Plate 9. Cameo glass vase by Emile Gallé, *c.* 1895. Gallé's work at the end of the nineteenth century showed a departure from the traditional nineteenth-century view of glass as a material which could be made to imitate other substances such as porcelain. He was interested in the possibilities that glass offered for new forms, colours and textures. His vase may be compared with that shown in Plate 10, an example of Stourbridge work.

Plate 10. Cameo vase, Stourbridge, *c.* 1885. The revival of the ancient art of cameo cutting in glass followed the work of John Northwood of Stourbridge who made a replica of the Portland Vase in 1873. He developed many new tools and techniques for carving, etching and engraving glass which were later used by the Stourbridge craftsmen to produce fine pieces for a wider public.

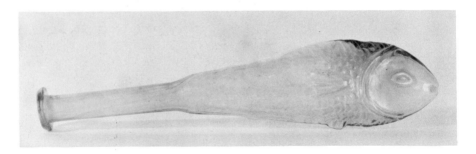

Plate 11. A Roman moulded 'fish' bottle, second to third centuries AD.

Plate 12. This nineteenth-century pressed glass tray with its elaborate ornamentation is typical of ware produced in the USA in imitation of the more expensive hand cut crystal glass. By developing the means of mass production and so drastically cutting the price of the product the American manufacturers were able for the first time to supply the general public with glassware for everyday use.

Plate 13. A free-blown 'pouch' bottle, seventh century AD. After the fall of the Roman Empire the styles of glass vessels became much simpler and less varied, and many techniques, such as mould blowing, fell into disuse.

Plate 14. The Lycurgus Cup. This picture shows a different view of the Cup to that seen in colour on the jacket. Here Dionysus abuses Lycurgus as he struggles in the coils of the vine.

to become of great importance in the development of the modern glass industry. Here there was an immediately available coal supply, good sand was plentiful and alkali was obtained locally from raw pot or pearl ash. Limestone, kelp (for alkali) and lead were available from various states.

The development of the pressed glass industry

As American workers were scarce and wages were much higher than in Europe means were soon sought to increase productivity. A small amount of pressed glassware had been made in England and Holland in the early years of the nineteenth century, but by the middle of the nineteenth century acceptable cheap pressed ware was produced in the USA on a massive scale for general consumption as a substitute for the expensive cut lead crystal glass. The process of hand pressing is shown in Figure 6; 7 and 8 illustrate the principles of this process.

The results of these American developments were described in a government report of 1884. The original American spelling is retained:

> . . . as usually practiced a metallic plunger is driven into a metallic mold into which molten glass has been placed by mechanical means; the glass taking the form of the mold upon its outer surfaces, while the inner is modelled by the plunger itself. The simplest form of mold is a flat slab of metal with slightly raised sides. For articles of some complexity molds are made in two or more divisions, hinged together (joint molds) and open outwards. The chief parts of the mold are termed the 'collar' and the 'base'. . . . The molten glass having been gathered and dropped into the mold, a sufficient quantity is cut off, the mold is pushed under the plunger and the long lever pushed down. The plunger enters the mold, the glass is pressed into all parts of the same, the plastic mass solidifies, the plunger is withdrawn, the mold opened, and the glass in the required form is withdrawn. If too much glass is cut off the article is too thick; if too little, it fails to fill the mold and the article is spoilt.

For a long time workers in the pressed glass industry were regarded by the rest of the glassmaking trade as unskilled operatives and their wages were lower. However, skill was certainly required in the making of pressed glass, in the gathering of the correct amount of glass and in keeping the moulds at the right operating temperature.

In their efforts to improve the pressing of glass the Americans made considerable advances in related processes. Furnaces were enlarged to hold more pots and the design of the furnaces improved. There was a steady growth of the glass industry west of the Allegheny Mountains where fine sand was available and the cheap coal gradually replaced the dwindling stocks of wood as fuel. As coal firing was adopted the flint (lead) glass furnaces tended to revert to the English cone pattern, although the window glass furnaces melting soda-lime glass remained of the German type. In 1860 the New England Glass Co. installed an underground feeding device by which the coal was forced up through the bottom of the fire. This was a great improvement on the existing method of carrying fuel constantly through the glasshouse and stoking through a firing door at the level of the glasshouse floor which harmed the pots and affected the glass quality.

The quantity and variety of moulds increased. Early moulds had been made by hand or on a lathe from wood; soapstone, cast iron or brass were

Fig. 6. Hand pressing of glass, 1849. The worker on the left of the picture is placing molten glass in the mould. The worker on the right is cutting this glass from the gathering iron prior to forming the article by operating the lever and lowering the plunger into the mould.

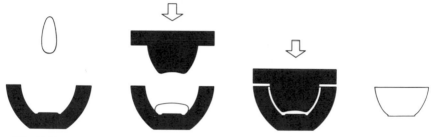

Fig. 7. Hand or automatic pressing: *left* to *right*: gob drops into mould; gob settles and plunger descends; final shape pressed; finished product.

Fig. 8. Manufacturing a vase by pressing. The plunger has just been withdrawn from the mould, which is now open to reveal the finished vase. On the right a closed mould can be seen.

used later but as brass was expensive a brass shell could be surrounded by iron. The first United States patent for an iron mould was taken out by Joseph Magoun in 1847 but the iron left the glass rough and was not really suitable as a mould material until 1866 when chilled iron moulds were introduced.

Up to about 1860 most pressed glass was flint (lead) glass and there was a great increase in flint glass production. Bottles which had previously been made in amber or green glass were also made in flint glass. Following the discovery of crude oil near Oil Creek, Pennsylvania in 1859, a cheap and plentiful supply of lamp oil was made available and there was a great expansion in the making of flint glass lamp chimneys; lampshade making was also very profitable in European countries.

The bottles, window glass and lampshades were always made out of soda-lime glass in the European countries, but lead glass was generally preferred by the Americans because they could make it of high quality. However, in 1864 W. Leighton, of the firm of J. H. Hobbs, Brockunier and Co., made a series of experiments on the composition of soda-lime glass. He is reported to have used sodium bicarbonate, but it is not clear how this was obtained, particularly as the statement antedates the widespread commercial use of the Solvay processs. His improvement of soda-lime glass was probably due to the use of purer ingredients and to a more careful choice of the proportions of the batch constituents. This high quality soda-lime glass was obtainable at less than half the cost of flint glass and was soon being used for the manufacture of most pressed glassware.

Soda-lime glass chills and becomes rigid more rapidly than flint glass and therefore faster forming methods were needed. Several mechanical improvements followed, such as the incorporation of springs into the press, the replacement of the straight wooden press lever by a more efficient weight-balanced curved steel lever, and various improvements to the moulds. Plungers quickly became overheated; this was remedied first by water cooling and then in 1886, a compressed air system was evolved. Hot glass was first used to bring the moulds to the right temperature but in 1860 mould-heating furnaces were introduced which saved time in the heating of moulds, saved glass, and resulted in uniform temperatures.

The piece of American pressed glass (plate 12 between pages 36 and 37) is typical of the products of this rapidly growing branch of the industry. Its expansion was part of a general growth of American industry and this period also saw the rise of effectively organized trade unions. In all sectors of the glass industry these unions developed a comprehensive system of regulating production that lasted into the twentieth century. Rates of production, apprenticeship schemes, wage rates, and the length of the working 'year'—the period from autumn to early summer when glassmaking was carried on—were all subjects of negotiation. All glass in the USA was therefore expensive, imported glass because of high protective tariffs and home produced ware because of high wages and limitation of output by the unions. The pressed glass industry was the remarkable exception. The pressed glass workers were regarded as unskilled by the unions, they were not so well organized and output limitation was not generally practised; increasing production resulting from technical innovation, relatively low wages and decreasing costs all combined to enable the USA to produce the best and cheapest pressed glass during the second half of the nineteenth century.

Although pressed glass was in the forefront of mechanization in the glass industry, subsequent changes were more gradual than those which took place upon the introduction of mechanization in container and window glass processes. Minor developments occurred, such as the placing of several moulds on a turntable, and similar arrangements were made for finishing pressed articles by heating with gas flames, but no great changes came until 1917 when automatic glass feeders, which fed lumps of glass of the correct size continuously from a tank furnace, were adopted for use with pressing machines.

The introduction of mechanical glass manufacturing processes

Container glass. In 1890 the American container industry was still basically a craft industry. Bottles were made by hand, the body of the bottle being blown first, followed by the forming or 'finishing' of the neck. Successful machine manufacture was only possible after the inventive step had been made of reversing this process so that the finish, or mouth part, of the bottle was formed first; the machine holding this end could then proceed to blow the body of the bottle. The sequences of operations for the blow-and-blow process for narrow mouth containers and the press-and-blow process for wide mouth containers are shown in schematic form in Figures 65 and 66 on pages 190 and 192 in Chapter 7.

In 1865 Gillinder patented a process for the making of pitchers by first pressing and then free-hand blowing. There were several attempts to make press-and-blow machines notably those of Arbogast and the Englishman Ashley in the 1880s, but until 1892 most American containers were made by hand. In 1893 the Enterprise Glass Company secured the Arbogast patent rights and started to make Vaseline jars on a commercial scale. A combined pressing and blowing mould, which made transfer between pressing and blowing operations unnecessary, was one of many improvements introduced and there was soon widespread production of wide-mouth ware such as pots for meat pastes and preserves. Less skill was required to operate the new machines and the original team was eventually reduced to two skilled and two unskilled workers.

The blow-and-blow process for the mechanical manufacture of narrow-mouthed objects was also developed at about this time, notably by Ashley in England and later in France, Germany and the USA. Both the press-and-blow and blow-and-blow processes were at this stage dependent upon an operator feeding the correct amount of glass to the forming machine and are therefore known as semi-automatic. Semi-automatic processes can be operated using glass from pots but once both feeding and forming are done automatically, a continuous controlled supply of glass is required, which was impossible to obtain until the advent of the regenerative tank furnace. The introduction of the first automatic glass-forming machine, the Owens bottle machine at the beginning of the twentieth century, and the continuous processes for making window glass had to await the advent of the tank furnace. The rapid adoption of the Siemens gas-fired regenerative tank furnace by the European countries during the late 1860s was not repeated in the USA; by 1880 coal was the established fuel of the industry but a few years later natural gas was discovered in the USA and the regenerative furnace came into its own. The glass industry was still on a small scale and the buildings were not very permanent structures; it was therefore easy to move to areas near the natural gas fields; in 1870 Ohio had nine glass factories but by 1890 there were sixty-seven. Natural gas was a very convenient clean fuel and by the end of the century the glass melting tank was in general use.

In automatic feeding and forming it is essential that at all stages of the process the glass has the correct viscosity, or stiffness; the viscosity varies very rapidly with the temperature which must be carefully controlled. Natural gas made it possible to exercise much greater control over furnace temperature than had been possible with wood or coal firing. Although the natural gas boom in Indiana, Ohio and Illinois was over by about 1910 improvements in producer-gas furnaces permitted the continuation of many glassworks in the old natural gas areas and by 1920 most plants were using producer gas, but some had already started to use oil firing. Gas firing continued to be widely practised and the extension of the long-distance transmission system from the gas wells of Texas has made natural gas the chief fuel in the USA.

In 1903 the Owens bottle machine began production (see Chapter 7). It had a suction device for feeding the correct quantity of glass to the mould and thus was the first fully automatic bottle-making machine. Production rates were generally increased, production costs dropped and with them the wages of the hand-blowers. The workers and their unions fought an increasingly desperate battle against the automatic machines and were forced to make one concession after another to the employers who adopted these machines. The Glass Bottle Blowers Union decided upon several new policies in the period 1903-9 in an effort to meet the competition of the Owens machine, the first being the introduction of a three-shift system in place of the traditional two-shift system, thus encouraging the employment of more blowers. Although the semi-automatic machines had at first been opposed, the Union now realized that they were preferable to automatic machines in their requirements for skilled manpower and so adopted a more favourable attitude towards them. They took advantage of the introduction of two- and one-man semi-automatic machines to secure higher piece rates for machine operation. Finally the Union continually reduced the piece rates in hand factories in order to compete with the output from the Owens machines at a time when the piece rates for the men working these machines were continually rising owing to increasing production efficiency. The wages of machine and hand-workers became equal in 1914 and by 1917 machine wages had risen above those of the hand-workers.

In 1917 the unions abandoned one of their last restrictions, the compulsory summer stop, and continuous production was therefore possible. After 1914 many patents were taken out for automatic feeding devices and alternatives to the Owens machine became available. By 1920 almost any bottle could be produced better on an automatic machine than by the hand processes of 1890.

Many other factors played an important part in increasing production and lowering costs. From 1903 batch mixing, conveying and charging were increasingly mechanized. The annealing of glass was improved by the gradual introduction after 1880 of the modern type of lehr, a long, tunnel-like annealing furnace in which the temperature was varied and controlled along its length so that the glass articles gradually cooled as they passed through. The word lehr, which is now in common use, first appeared near the end of the nineteenth century in the USA. It derives from the older forms 'leer' or 'lear'; leer seems to have been used first during the seventeenth century by Merrett to describe the heated oven in which the glass was annealed.

By 1912 the American container industry led the world in bottle and jar production. With the development of automatic feeders the importance of the semi-automatic machines declined after 1917 while the Owens machine was replaced by feeder-fed machines during the 1920s. The decline was aided by the high installation costs and the restrictive licensing policy of the owners of the Owens machine patents. In twenty-five years the American bottle industry had changed from one in which the great majority of ware was made by hand, to an industry producing over ninety per cent of its output by automatic processes and exporting over $3,000,000 worth of containers annually. In 1901 the total production of containers was 12,005 thousand gross and by 1925 it had increased to 26,044 thousand gross. These figures may be compared with those for 1970 of 267,179 thousand gross.

Window glass. In 1890 there were few tank furnaces in the USA and the window glass industry had reverted to the older habits of following fuel supplies; the fuel was now natural gas which had rapidly replaced coal, but by 1900 the window and cast glass industry had 353 natural-gas fired tank furnaces in operation with a capacity equivalent to that of 4834 pots. Although producing good window glass and using many new innovations, including automatic delivery and mixing of batch materials, automatic charging and transfer of glass on conveyor belts to an improved annealing lehr, the new factories still employed the traditional forming methods.

In 1903 a machine for drawing large cylinders of glass upwards from a pot of molten glass was developed by John H. Lubbers, an American window glass flattener, who had started his experiments in 1896. The process was still intermittent and the cylinders had to be cut and flattened but much larger cylinders could be made much more quickly than by hand and the team of workers, although larger than before, required less skill; the skilled blower and gatherer were dispensed with.

This invention, as with all new processes, brought about strife and competition between the various interests in the industry and the competition was intensified by the introduction of the Colburn machine for the continuous mechanical drawing of a sheet of glass from molten glass in the tank. The Colburn Machine Glass Company was formed in 1906 but sheet-drawn glass was not made commercially until 1917. The advantages over hand and machine cylinder glass were very great: the process was continuous with less wasted time and labour, the glass was free from flattening faults with uniform thickness and a natural fire polish.

Temporarily increased demand following the disruption during the First World War of foreign glass industries which had previously shipped large quantities of glass to the United States at competitive prices, enabled hand-made cylinder glass to survive a little longer in face of the competition from automatic machines. However, the Colburn process was widely adopted in the United States during the 1920s and window glass could now be made more efficiently and more cheaply than either by the Lubbers cylinder process or by hand manufacture. In 1920 there were 329 cylinder machines, 6 sheet machines and 2367 hand pots; by 1929 there were 60 cylinder machines, 117 sheet machines and no hand workers. The Colburn machine eventually lost ground to the Belgian Fourcault process which produced superior glass because the glass moved vertically upwards, instead of being bent over to pass through a horizontal annealing oven as in the Colburn process.

The major development in the plate glass industry during these years was the introduction of the continuous lehr. Plate glass manufacture as a commercially successful venture had been developed by J. B. Ford on a very large scale in the 1890s. Although by 1890 the domestic industry was supplying most of the country's requirements, traditional equipment used in plate glass manufacture was generally imported from England, such as rectangular casting tables which had to be moved from one annealing furnace to another. During the early 1900s a continuous annealing lehr, which had a runway 300 feet long, was introduced. The cast plates were passed through the lehr slowly from a stationary casting table and the annealing process could be completed in three hours instead of the forty-eight hours previously required. Power-lifting apparatus and improvements in grinding and polishing methods were introduced and in the 1920s a continuous method for the manufacture of plate glass was developed.

Although the adoption of automatic methods for sheet glass manufacture was completed during the 1920s the American window glass industry was not in such a strong competitive position in world trade as the container industry. Window glass was still imported in spite of high tariffs. Wages were lower in Europe and the ocean freight rates were so low that the European glass industry could compete successfully for the coastal markets of the USA with the home industry, the products of which were transported for long distances at high cost overland. Small panes were imported because the American workers could earn more money by making the larger sizes. After the First World War there were great advances in glassmaking in Europe, particularly in Belgium, and foreign syndicates were formed which became serious competitors to the American window glass industry.

Lamp bulbs. The Lubbers machine mechanized and improved the hand operations of the cylinder process of making sheet glass, but the Colburn and Fourcault processes were completely new ideas in which the sheet was drawn continuously from the tank. The automatic production of bulbs for electric lamps developed in a similar way, the procedures of the hand blower being mechanized on the Westlake machine. In the hand process bulbs were blown in wetted wooden moulds; rotation of the blowing iron during the blowing-up of the bulb produced a high polish and a bulb with no mould marks on its surface. Later improved moulds of cast iron were coated with a mixture of linseed oil and charcoal or hardwood sawdust and heated to carbonize the paste. These paste-moulds were still dipped in water before each blowing to permit easier rotation. In 1879 the Corning Glass works were making glass bulbs for Edison who brought out his first incandescent electric lamps in 1881. The electric lamp bulb industry grew rapidly and provided the stimulus for the development of mechanical manufacturing processes.

The Owens semi-automatic process of 1894-5 was essentially the original hand-gathering but the blank was rotated and shaped mechanically, allowed to elongate and again rotated whilst the bulb was blown inside the blow mould using compressed air. Although the process required no skilled blower and the rate of production was much higher than by the hand method, a gatherer and three unskilled workers were still needed.

During the years 1910-20 two completely automatic bulb-blowing machines, the Empire and the Westlake, were introduced. The Empire machine was first made in 1915 but in about 1923 an automatic feeder enabled gobs—lumps of glass of the correct size—to be fed alternately to two blank moulds situated on intermittently rotating tables. The blanks were pressed to form the initial shape or parison, reheated, allowed to elongate under their own weight and blown to their final shape in the blow mould, each operation occurring after rotation of the table holding the forming units. The formed bulb was ejected and the rough edges cut off and fire-polished—a completely automatic process. The machine was powered by electricity.

The Westlake machine, introduced in about 1916, combined suction-gathering and paste-mould forming. The machine had two gathering arms which entered the furnace chamber and sucked up a sufficient amount of glass. The charge of glass was dropped into a shallow cup on the end of an inverted, slowly turning blowpipe, twelve of which were mounted on a rotating centre. The blowpipe then swung over with the glass hanging from its end. The bulb was blown to its final shape in a wetted paste-mould which rose from a water trough to enclose the glass, thus imitating the action of the hand blowpipe. These bulb machines, suction or feeder fed, carried the displacement of workers a stage further; gatherers, blowers and knockers-off, the boys who cracked the bulbs off the blowing irons and took them to the lehr, were no longer required.

The Ohio machine which followed the Westlake had arms which had shrunk to quite short spindles and the actions of the hand-blower could no longer be so easily recognized. Finally, in the Corning ribbon machine (Figure 9), developed during the late 1920s and used for modern lamp bulb production, a ribbon of glass formed by passing the glass through water-cooled rollers moves under a line of moving blow-heads; the partially formed bulbs still attached to the blow-heads are then enclosed in a line of blow-moulds moving below the blow-head line and when the final bulb shape has been blown the moulds open and the bulbs are automatically cracked off and carried away on a conveyor belt. Thus the link with the hand-blower has now completely disappeared and one of these ribbon machines can make over 2000 bulbs a minute. The first ribbon machine outside the USA was installed in England at Glass Bulbs Ltd., of Harworth, Yorkshire, in 1950; it was said that it could make enough bulbs for all Europe; its complexity and capacity for mass production, however, still mean that it can only be installed in countries with advanced technologies and economies.

Glass tubing. Until 1917 glass tubing was made by a hand process (Figure 10). A suitable quantity of molten glass was blown and marvered into a short hollow cylinder with thick sides. An iron rod was then attached at the end opposite to the blowpipe to the base of the cylinder, which was then rapidly stretched out to form a narrow tube; one worker walked away from the other as they held the rod and the blowpipe. The tube often showed a tendency to thin at various points and a third worker would fan the glass at these points so that it became rigid and was prevented from further narrowing. Even so, the wastage was very high, only about twenty-five per cent of the glass being sold. The bore of the tubes was not uniform and there was a high breakage rate owing to unequal cooling of the glass, but a mechanized version of this hand process is still used for drawing thermometer tubing.

In 1917 Edward Danner at the Libbey Glass Company introduced an automatic method for tube making which dispensed with skilled hand-workers; he had worked on the problem for years and had finally solved it during a long holiday on which he had been sent to recover from overwork. Glass was allowed to flow continuously from a hole in the melting furnace onto a slowly revolving refractory mandrel which it completely covered. The glass was drawn from the end of the mandrel and pulled onto a series of rollers on which it travelled until it was cool enough to be cut off into

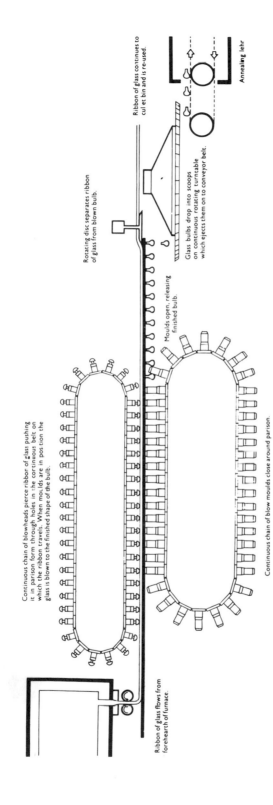

Continuous chain of blowheads pierce ribbon of glass pushing it in parison form through holes in the continuous belt on which the ribbon travels. When moulds are in position the glass is blown to the finished shape of the bulb.

Rotating disc separates ribbon of glass from blown bulb.

Moulds open, releasing finished bulb.

Continuous chain of blow moulds close around parison.

Ribbon of glass flows from forehearth of furnace.

Ribbon of glass continues to cullet bin and is re-used.

Glass bulbs drop into scoops on continuous rotating turntable which ejects them on to conveyor belt.

Annealing lehr

Fig. 9. The ribbon machine for the automatic production of light bulbs. One of these machines can make over 2000 bulbs a minute.

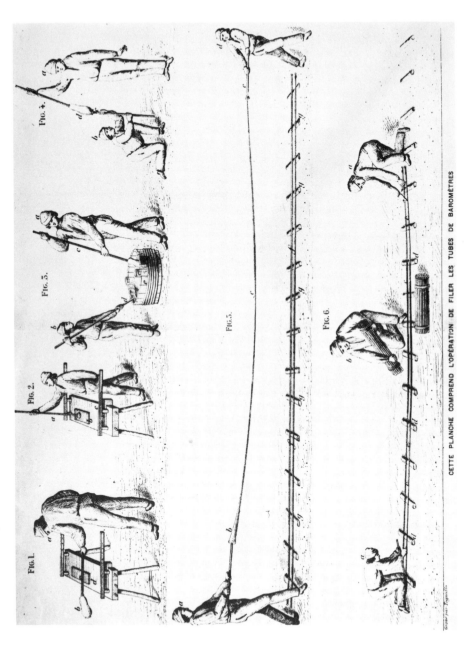

CETTE PLANCHE COMPREND L'OPÉRATION DE FILER LES TUBES DE BAROMÈTRES

Fig. 10. This nineteenth-century drawing shows the hand manufacture of glass tubing. In the top line the initial gather of glass is taken on the blowing iron, roughly shaped into a short hollow cylinder by blowing, and then attached at its other end to an iron rod. In the second line the workmen are forming the tube by rapidly walking away from each other whilst holding the rod and the iron, and in the bottom line the tube is being cut up into lengths.

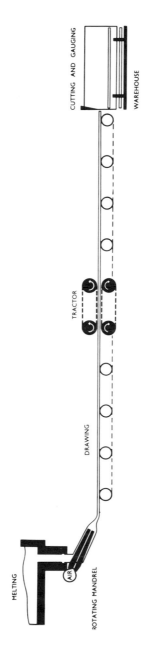

Fig. 11. The automatic production of glass tubing by the Danner process: schematic arrangement.

Fig. 12. View into the furnace showing the molten glass flowing onto the mandrel and being drawn off as a continuous tube.

lengths; this process produced solid rods. For tubes, compressed air was continuously supplied through a small pipe to the centre of the mandrel as the glass was drawn off. The Danner process considerably reduced the cost of making tubing and in the same time one machine could produce about the same amount as ninety workers. Figures 11 and 12 show how glass tubing is made automatically on this machine.

Thus, in the manufacture of tubing as in the container, window, pressed and lamp glass industries, the nineteenth and early twentieth centuries saw a revolution in glassmaking methods in the USA. From being an industry based on ancient craft methods almost all aspects of the glass industry had become outstanding examples of modern mechanized production.

Bibliography

1. *The Development of the American Glass Industry,* P. Davis, Harvard University Press, 1949.
2. *Revolution in Glass-making,* W. C. Scoville, Harvard University Press, 1948.
3. *La verrerie au XX siècle,* J. Henrivaux, Geisler, 1911.
4. *The British Glass Industry: its development and its outlook,* W. E. S. Turner, *J. Soc. Glass. Technol.,* 1922, **6,** 108.
5. *Pilkington Brothers and the Glass Industry,* T. C. Barker, George Allen and Unwin Ltd., 1960.
6. *A History of Chance Brothers and Co.,* J. F. Chance, privately printed by Spottiswoode, Ballantyne and Co. Ltd., 1919.

3

Ancient glass compositions

The chemical compositions of modern glasses are carefully controlled to produce materials with suitable physical and chemical properties for specific applications. In addition to the familiar soda-lime-silica glass, lead crystal, and the borosilicate glass of low thermal expansion used for chemical apparatus and heat-resistant cooking ware, many special compositions are made for street lighting lamps, for fibreglass, for several applications in the electrical industry, and particularly for optical purposes. The development of new glass compositions for specific applications followed the advance of chemical knowledge during the nineteenth century and the demands of new technologies in the present century, but for the first few thousand years of the craft, glassmaking was essentially the heating of sand with a flux such as plant ashes.

Specially selected rocks, sand or pebbles have always been the source of the silica. Today sand from special deposits is often washed and treated chemically before use. Pure crystalline silica melts at 1723°C, but even at the melting temperature it is a very viscous liquid, its coefficient of viscosity is 10^5 poises, which means that it is about ten million times more viscous than water, or one thousand times less pourable than syrup. It is far too stiff for normal glassworking methods and the high melting temperature is very difficult and costly to attain. The addition of soda 'fluxes' the silica and can lower the melting point to below 800°C, but a glass of this low melting composition would be very easily attacked by water. The addition of lime increases the durability, *i.e.* the resistance to attack by water; however, too much lime gives a glass which is prone to crystallization. The successful ancient glass compositions must have been found by long experience and careful selection of sand and fluxes. These compositions almost inevitably were not very different from modern soda-lime-silica glasses, for the range of mixtures which make satisfactory glass is not very large.

To the selected sand and pebbles the early glassmakers added as the flux, plant ash or alkaline salts from natural deposits. Lime was very rarely added deliberately but was always present in the sand or ash. Every chemical analysis has revealed the presence of many other components, some of which were introduced deliberately in an attempt to imitate precious or semi-precious stones; but many came from impurities in batch materials, melting pots and fuels. This random complexity has made it very difficult to use analysis to date a sample of ancient glass or to specify its place of origin; nevertheless modern analytical techniques are revealing a great deal about the

52

technology and skills of ancient times. Klaproth, at the end of the eighteenth century, made the very first analyses of glasses and, until about 1950, quantitative analysis was made using traditional chemical methods in which the chief measuring instruments were the burette, pipette and balance. In the 1930s, X-ray crystallography and spectroscopy became invaluable aids and during the last few decades many additional tools have become available to the analysts, particularly spectrometers of various kinds and lately X-ray fluorescence spectroscopy. In this latter technique a specimen is bombarded with X-rays or electrons which cause the various elements present to emit their own characteristic fluorescent X-radiation.

The analysis of the Lycurgus Cup

The power of these modern techniques has been demonstrated in a recent investigation of the Lycurgus Cup, a notable specimen of Roman glass dating from approximately 400 AD (see plate 14). This Cup is a remarkable example of glass cutting and is of great interest for its colours. The colour of the glass when seen by transmitted light is a deep magenta, whilst by reflected light it is a greenish gold or jade; this dichroism is extremely rare amongst ancient glasses. The chemistry can now be readily explained but as with the Portland Vase the difficult technical problems must have been solved by years of careful observation and experimentation.

A tiny chip of glass was found dislodged from the foot of the Cup when the modern silver mounting was removed and on this a complete analysis has been made. X-ray diffraction showed that there was no detectable amount of crystalline matter present, and spectrographic examination revealed that the elements silicon, sodium, potassium, calcium, magnesium, aluminium, boron, manganese, were present in significant proportions, with the minor constituents, iron, titanium, silver, gold, antimony, lead, tin, barium, strontium. After these tests, the eight major constituents were determined quantitatively by special techniques: only seven milligrams of glass were used in the determination. The Cup was the first example of ancient glass found to contain gold and silver. As there were insufficient amounts of these elements present in the original sample for direct determination, experimental glasses were prepared of the same general composition as that of the Lycurgus Cup, containing varying amounts of gold. Direct spectrographic comparison then revealed that the gold content of the Cup is in the range 0.003 to 0.005 per cent.

Other soda-lime glasses were also melted containing small amounts of one or more of the elements gold, silver, antimony, tin and manganese, because it had previously been suggested that the unusual colouring of the Cup could have been caused by combinations of these elements. The glasses were used as chemically calibrated standards for X-ray fluorescence spectrometry. From a comparison of the X-ray intensities of the Cup material and the calibrated standards, the silver, iron and antimony contents were estimated. Because the fluorescent rays came from all parts of the specimen, the shape and size of the reference samples had to be chosen to match as closely as possible those of the original Cup sample in order to avoid errors arising because of re-absorption of the characteristic X-rays by the specimen itself. Lead and copper were also determined by X-ray fluorescence spectrometry. An ancient glass in which the content of lead had already been determined was used as a reference standard for lead, and the copper was estimated indirectly by relating the signal to the observed intensity from the iron. All these investigations had consumed only seventy-eight milligrams of the chip and there was sufficient remaining for a direct determination to be made of silver and phosphorus by chemical methods.

Thus, with the expenditure of only a few milligrams of glass, fourteen elements had been accurately determined. Although the minute quantities of gold and silver were responsible for the dichroic 'Lycurgus effect', simply adding the traces of gold and silver to the glass would not ensure that the delicate shades of green and red would be obtained. Only by using reducing agents could the gold and silver be chemically reduced and only by appropriate heat treatment could submicroscopic crystals of the metals be made to grow in the glass. During this heat treatment the crystals grow through a range of sizes with resulting changes in absorption and scattering of light, and hence in the colour of the glass, which depends quite sensitively upon the proportions and oxidation states of certain elements within the glass, and upon the time and temperature of heating while the colour is developing. The Cup is thus remarkable not only for its fine cutting, but for the patience and skill which led to the final achievement at a time when there was no chemical science on which to base an understanding of the processes involved.

Coloured glasses and opalescence

Gold and silver were not the only colouring agents known in the ancient world. A whole range of blue glasses was obtained from copper and cobalt, alone or in combination, and with iron and manganese. Greens were obtained by iron and by copper, purples by manganese, opaque red by cuprous oxide, white opal by antimony, and opaque yellow by antimony in combination with lead.

Until recent years the view was widely held that ancient opal glasses were produced by the inclusion of tin oxide in the glassmaking mixture, but this view was based on very slender analytical evidence. An extensive study of the materials used by the ancient glass-workers for producing white opal and opaque glasses in various shades of blue, green, turquoise, red and yellow has shown that antimony oxide was widely used. A small piece of the opal glass overlay of the Portland Vase was examined by X-ray diffraction; antimony rather than tin oxide was shown to be the crystalline substance producing the opalescence. This discovery led to the examination of thirty-seven Western glasses, from the earliest available specimens to glasses made in modern times. Down to the first or second centuries AD, antimony oxide was the dominant material in the production of opal or opaque coloured glasses other than in some red opaque glasses, in which cuprous oxide, Cu_2O, or cuprous oxide plus metallic copper was found. During the early Christian era other opal producing substances came into use in the West including tin oxide and calcium phosphate. In the eighteenth century the intense opal produced by precipitates of lead arsenate was discovered and fluorides were introduced in the second half of the nineteenth century. However, it is known from X-ray diffraction studies that fluoride opal was being made in China in the seventh century AD, preceding its production in the West by many centuries.

Many other Western glass samples have been examined. It now appears that at some time between the second and the fifth centuries AD, antimony oxide ceased, for unknown reasons, to be employed as the sole opacifying agent, and its place was taken by tin oxide. It is difficult to suggest any reasons for this change as deposits of workable tin and antimony ores occurred in various parts of Europe and both metals had been isolated in very early times and used widely in materials such as bronzes. Nor was there any particular difficulty in mining the tin ore cassiterite and stibnite, the ore of antimony.

However, the separation and identification of the various metals was not possible in ancient times, tin and lead were confused and the first use of tin oxide as an opacifying agent may have been accidental. Tin was known as white lead and no way was known of separating tin and lead when they occurred together in the ore. A rough test of the proportions of lead and tin in the alloy could be performed by pouring it onto papyrus: melted tin would not scorch the papyrus whilst molten lead would do so. Antimony as stibnite and lead compounds were also confused, an example being the early Egyptian eye-paint known as stibium, but in fact composed of galena, a lead compound. The ancients also referred to both antimony and its ore as 'stimmi', or even as lead.

The composition of ancient glasses

Apart from the small quantities of materials added to produce colour or opalescence, recipes for making glasses free from lead (the vast majority of glasses of the Western ancient world) from 700 BC to the seventeenth century all prescribed crushed silica rock or sand and ash (or 'glassmakers' salts') as the major constituents. Yet all analyses disclose the presence in the ancient glass of from two to three per cent up to more than twenty per cent of lime, 0.2 per cent to seven per cent of magnesia, and small amounts of alumina and other oxides. These constituents were derived from the major batch materials or from the corrosion of the crucible.

Sands

The locations of two sands for glassmaking are specifically mentioned in classical writings. One was at the mouth of the River Belus on the Syrian coast and the other was a seashore deposit mentioned by Pliny, near the mouth of the River Volturnus north-west of Naples. The Belus sand enjoyed a good reputation for many centuries as it served the glassmakers of the Syrian coast and was also widely exported. This sand has been found to contain lime equivalent to 14.5 to 18 per cent of calcium carbonate, 3.5 to 5.2 per cent of alumina, and about 1.5 per cent of magnesium carbonate; when combined with a suitable amount of sodium carbonate it would make an excellent durable glass. Moreover, the iron oxide content, 0.12 to 0.15 per cent, is low in comparison with Egyptian sands which were probably used for glassmaking. It might just be used for modern window glass, but for colourless bottles the iron content of the sand used today is generally below 0.05 per cent.

The sources of sand employed in ancient Egypt are unknown. Most Egyptian glasses contain substantial amounts of iron oxide and it is almost certain that only the impure desert sands could have been used although selection of desert sands should have provided some with low iron contents; these sands also contain the necessary lime and alumina. Purity of the sand was always a major concern of the glassmakers: Merrett in 1662 in discussing the sources of glass used in England distinguished between 'fine white sand from Maidstone in Kent' for high quality glassware, and 'a coarser sand from Woolwich for green glasses'.

Alkalis

At least four sources of alkaline salts were available to the ancient glassmaker; natural deposits resulting from evaporation and drying up of former land-locked seas and lakes, leaching of salts from soils, deliberate evaporation of sea or river water in pans or pits, and the ashes of vegetable matter.

The natron from the Wadi Natrun in Egypt was certainly used for glassmaking. This material contains the carbonate and bicarbonate of sodium, and was used from earliest times for a variety of purposes: as a detergent, in medicine, and in the process of embalming. Both ancient and modern samples show complex and widely varying compositions, but ancient Egyptian glasses have been found to contain much more potash than can be derived from natron or natural soda. The composition of modern natron from the Wadi Natrun is given in Table 1.

Salts of various compositions were obtained from lakes or rivers by evaporation. Figure 13 shows a modern source of sodium carbonate produced by evaporation, Lake Magadi in Kenya. According to Pliny, the Nile waters were employed in the preparation of 'nitrum', the water being passed through sluices and channels into shallow pits where it evaporated and yielded 'salt'. A

Table 1. Modern natron from the Wadi Natrun

	%	%	%	%	%	%	%	%	%	%	%	%	%	%
Sodium carbonate	38.2	22.4	28.9	35.5	43.5	28.9	58.6	75.0	67.8	33.4	38.3	41.8	35.4	53.9
Sodium bicarbonate	32.4	6.2	20.5	25.8	33.8	9.9	14.3	5.0	8.6	25.2	18.3	29.4	12.1	24.2
Sodium chloride	6.7	26.4	24.8	14.0	4.8	26.8	7.4	9.4	4.3	20.8	2.2	11.9	12.4	1.9
Sodium sulphate	2.3	39.3	5.8	3.0	3.3	27.4	1.3	1.2	0.8	6.1	tr.	3.4	29.9	tr.
Water (free and combined)	16.5	5.6	12.8	13.1	13.1	6.9	4.3	3.7	1.9	11.6	10.1	11.2	10.2	20.0
Matter insoluble in water	3.9	0.1	7.2	8.6	1.5	0.1	14.1	5.7	16.6	2.9	31.1	2.3	tr.	tr.
	100.0	100.0	100.0	100.0	100.0	100.0	100.0	100.0	100.0	100.0	100.0	100.0	100.0	100.0

Fig. 13. Trona (sodium sesquicarbonate—$Na_2CO_3.NaHCO_3.2H_2O$) in Lake Magadi, Kenya. The crust of the disturbed material is two to three inches thick, but underneath the crust the lake is still solid with saturated solution (liquor) filling the interstices between the crystals. As the solid material is dredged out a pool of liquor forms which supports the floating dredge but which in time crystallizes out again forming fresh trona. In the rainy season the surface of the lake becomes flooded to a depth of a few inches but underneath the material remains solid and dredging continues.

recent analysis of the matter in solution in Nile water showed that sodium, calcium and magnesium carbonates are present in approximately equal quantities as the main constituents. Potash is present but soda is the predominant alkali. The task of preparing the salt would be quite formidable as 7000 parts of water would have to be evaporated in order to yield a single part of salt. Whether or not the nitrum was so prepared, after the subsidence of the floods of the Nile the evaporation of the water certainly left the ground richer in these glassmaking materials. Although nitrum was the word used by Pliny to describe the salt obtained from the Nile water, it was also used for the plant ash obtained from the oak tree. The latter consisted mainly of potash and the former of soda but as the distinction between the two forms of alkali was not established until the eighteenth century, alkaline salts were described under a general heading deriving from the Arabic word, *natrûn*.

Plant ashes also have a long tradition in glassmaking. In one of the earliest records of glass recipes, the Assyrian clay tablets from the Royal Library of Assur-bani-pal (668-626 BC), the alkali is derived from salicornia, a plant containing soda which is very common in the Near East.

A considerable number of glass recipes were recorded but many are unfortunately lacking in the name or quantity of some ingredient. The most complete and clearest of the recipes is for sirşu-glass:

> To make sirşu-glass: 20 mana of sand, 60 mana of salicornia-alkali, $1\frac{2}{3}$ mana of saltpetre, $\frac{2}{3}$ mana of lime.

Note the deliberate addition of lime, which is most unusual in the old glass texts. Figure 14 shows the tablet; the righthand column is a copy of the upper half, but includes additions from duplicate tablets.

Other ancient formulae for glassmaking show great variations in the quantities of ingredients prescribed. Pliny in the first century AD says:

> ... white sand ... is prepared for use by pounding it with a pestle and mortar which done it is mixed with three parts of nitre, either by weight or measure, and when fused is transferred to another furnace.

Nitre is here used as a general term for the alkaline salt.

In the twelfth century AD the German monk Theophilus prescribed:

> two parts of the ashes of which we have spoken (beechwood ashes) and a third of sand carefully purified from earth and stones, which sand you shall have taken out of water, mix them together in a clean place.

A similar mixture was described four hundred years later by Biringuccio, but with different proportions of the constituents. In his book of 1540, *Pirotechnia*, he describes

> The method of composing glass. ... First one takes ashes made from the saltwort that comes from Syria. ... Now some say that

Fig. 14. Seventh century BC Assyrian clay tablet, one of a series of glass texts made for Assur-bani-pal and devoted by him to the Temple of Nabû in Nineveh. The texts contain recipes for many types of glass and also instructions for making furnaces; they stress the importance of ceremonial purity in the work and the placation of the spirits, so that the glassmaking may be brought to a successful conclusion.

this ash is made from fern and some from lichen; which of these it does not matter here . . . some of those sparkling white river stones that are called pebbles and that are clear and breakable and have a certain resemblance to glass When it is impossible to have these, take in their place a certain white mine sand that has a certain rough harshness. Of whichever of these is taken, two parts are put to one of the said salt (ashes) and a certain quantity of manganese according to your discretion.

Variability in composition is not confined to that between different kinds of plant; it also arises from the growing of plants on soils of differing types. Some examples of the varying compositions of plant ashes are given in Table 2.

The influence of Neri

The textbook *L'Arte Vetraria* published by Neri in 1612 has been mentioned in Chapter 1. It was the first systematic account of the preparation of the raw materials for glassmaking, and it also contained recipes and melting instructions for many coloured and clear glasses. This book had a tremendous influence on European glassmaking: it was translated into English, Latin, German, Spanish and French, many of the authors adding extensive sections on glassmaking in their own countries. It was still in use at least until the end of the eighteenth century, as is shown in the third edition of the Encyclopaedia Britannica (1797) where the author recommends Neri's recipe for making 'crystal glass'.

Very little is known about Neri, except that he was a practical glassmaker who carried out much of his work in Pisa, but he also seems to have worked in Antwerp, Florence and Flanders. His approach to glassmaking was similar to that of the ancient glassmakers but although he still used the old names for materials he often showed some understanding of their chemical nature. For example, in talking of *crocus martis* (iron oxide) he said: 'Crocus martis is nothing else but a . . . calcination (product of heating) of iron.'

Neri realized that high quality glass could only be made from the purest materials available. His crystal glass was distinguished from ordinary green glass by the fact that the materials for the crystal glass were subject to a much more rigorous selection and purification process. Only certain types of rock or sand were suitable as a source of silica. Neri recommended that the whitest pebbles be chosen 'which hath not black veins, nor yellowish like rust in them . . . those stones which strike fire with a steel are fit to vitrifie and to make . . . crystall'.

When the selected silica had been reduced to a fine powder and sieved and the purified alkali obtained by a process of lixiviation, as described in Chapter 1, the two had to be combined in the correct proportions, which varied according to the alkalinity of the ash. Neri advised that these

Table 2. Percentage ash content and composition of various air-dried vegetable materials (calculated from data in Thorpe's *Dictionary of Applied Chemistry*, 1937, Vol. 1, pp. 508-9)

Substance	% Total ash	% Composition of ash								
		SiO_2	CaO	MgO	Na_2O	K_2O	P_2O_5	SO_3	S	Cl
Wheat straw	4.26	66.2	6.1	2.5	2.8	11.5	5.4	2.8	3.8	—
Barley straw	4.39	53.8	7.5	2.5	4.6	21.2	4.3	3.6	2.9	—
Heather	3.61	35.2	18.8	8.3	5.3	13.3	5.0	4.4	—	2.2
Fern	5.89	6.1	14.1	7.6	4.6	42.8	9.7	5.1	—	10.2
Reeds	3.85	71.4	6.0	1.3	0.26	8.6	2.1	2.8	—	—
Sedge	6.95	31.4	5.3	4.2	7.3	33.2	6.7	3.3	—	5.6
Rush	4.56	11.0	9.4	6.3	6.6	36.6	6.3	8.8	—	14.2
Wheat (grain)	1.77	1.7	3.3	12.4	3.3	31.1	46.3	2.2	—	8.4
Beech (leaves)	3.05	33.8	44.9	5.9	0.7	5.2	4.7	3.6	—	0.3
Beechwood, trunk	0.55	5.4	56.4	10.9	3.6	16.4	5.4	1.8	—	—
Beech brushwood	1.23	9.8	48.0	10.6	2.4	13.8	12.2	0.8	—	—
Oak	0.51	2.0	72.5	3.9	3.9	9.5	5.8	2.0	—	—
Apple	1.10	2.7	70.9	5.5	1.9	11.8	4.5	2.7	—	—
Mulberry	1.37	3.6	57.0	5.8	6.6	36.6	6.3	8.8	—	14.2

proportions could be determined by a series of experiments in which varying amounts of silica and ash were melted in small pots until a satisfactory glass was obtained. The two materials could then be combined in larger quantities and heated at a low temperature to form frit, which was then used as a basis for different types of glass.

Although Neri recognized the importance of pure materials and the necessity for combining them in the correct proportions, he did not realize that purification of the alkali removed lime and magnesia, the main stabilizing ingredients which provided resistance to weathering. The effects of the purification were noted by Merrett in his annotated English translation of Neri, 1662: '. . . in the finest glasses, wherein the salt is most purified, and in a greater proportion of salt to the sand, you shall find that such glasses standing long in subterraneous and moist places will fall to pieces, the union of the salt and sand decaying'. Both Merrett and Neri appear to have known of glass containing 'salt of lime', calcium oxide, and Merrett mentions unsuccessful experiments with 'lime' ground from oysters, crabs and lobsters. However, the understanding of the chemical function of lime had to await the work of Deslandes at the end of the eighteenth century, and the lack of stabilizing ingredients did not become critical until pure artificial soda was used on a large scale during the early nineteenth century (see Chapter 2).

Developments during the eighteenth century

During the eighteenth century glassmakers were making careful experiments and recording their results with the aim of finding new and successful compositions. For example, an English glass recipe book of 1778-9 (Figure 15) contains a series of experiments on the production of fine lead crystal glass, especially gold ruby glass. The recipes mention many of the raw materials which were in use at that time. A typical clear flint glass batch contained:

    20 lb of pearl ashes (purified potash made from wood ashes);
    70 lb of red lead;
    60 lb of washed Lynn sand;
    ½ oz of magnees (manganese dioxide); ,
    4 oz of arsnike (arsenious oxide);
and a gold ruby glass was made from:
    4 lb of salt petre;
    9lb 8 oz of Isle of Wight sand;
    9lb 8 oz of red lead;
    4½ drams of gold dissolved (in aqua regia);
    and precipitated with tin (*i.e.* Purple of Cassius).

The pearl ashes were being used in increasing quantities in place of ashes from the Levant and barilla from Spain, but a great variety of fluxes was

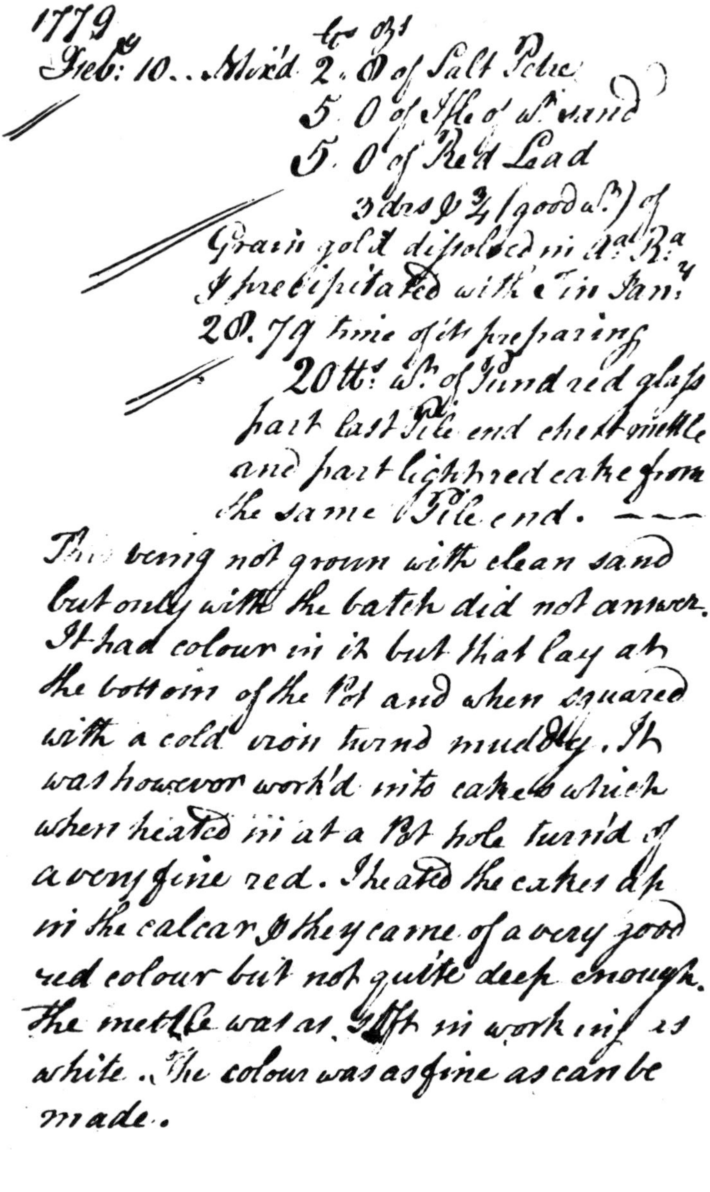

1779
Feb: 10... Mix'd 2. ℔ of Salt Petre
5. ℔ of Isle o' W^t sand
5. ℔ of Red Lead
3 drs & ¾ (good a^t.) of
Grain gold dissolved in a^q: R^a
& precipitated with Tin Jan^y
20. 79 time of its preparing
20 ℔: W^t of fine'd red glass
part last Tile end chest mettle
and part light red cake from
the same Tile end. ———
This being not grown with clean sand
but only with the batch did not answer.
It had colour in it but that lay at
the bottom of the Pot and when squared
with a cold iron turn'd muddy. It
was however work'd into cakes which
when heated in at a Pot hole turn'd of
a very fine red. I heated the cakes up
in the calcar & they came of a very good
red colour but not quite deep enough.
The mettle was as soft in working as
white. The colour was as fine as can be
made.

Fig. 15. 'Bradshaw's Recipes'—two pages from the notebook of an eighteenth-century glassmaker. These careful notes record experiments in the production of various types of glass, chiefly a 'gold ruby' glass, that is, a ruby glass in which the colouring effect is

1779

Feby 13.. Put 3 drams of B^d Gold into one Tumbler & 3 d^o. into another & dissolved them in Aqua Regia as before. The solution of one Tumbler was as much as an oz viol will hold & the other about the same quantity. I drop'd forty drops of the solution of one Tumbler into about ⅔ of a Pint of water & filed in filings of Tin which united with the solution & turned the water of a Pink or lightish red colour. when this was thoroughly mixed I drop'd in thirty or forty drops more & filed in more Tin filings I repeated the process till the red colour disapeared & the water became of a blackish olive colour from the quantity of solution in it; I then ~~continued to~~ filed in Tin till the water became purple

caused by a suspension of tiny gold particles. The glass described here contained saltpetre (potassium nitrate), Isle of Wight sand, red lead, and 'grain gold dissolved in aqua regia and precipitated with tin'.

used—calcined lead, a variety of plant ashes, sea-salt, borax, kelp, arsenic, clinkers from furnaces and wood ashes—in order to supply the glassmakers' increasing demands for alkali. Borax was highly valued as a very strong flux but as it had to be imported at high cost from the East Indies it was only used in high quality glass such as plate glass.

The action of manganese as a decolorizer was well known to Neri—'manganese consumes the natural greeness of glass'—and it was increasingly used following the publication of his book. English glassmakers could obtain high quality manganese dioxide from the Mendip Hills, where it occurred as a by-product of the lead mines.

Salt-petre had been used as a flux at least since the seventeenth century, but it also acted as a decolorizer in lead glass and as such is mentioned in many eighteenth-century recipes. 'Nitre . . . destroys . . . the phlogiston which gives a strong tinge of yellow to glass prepared with lead as flux, it serves to free it from this coloured tinge'.

Washed sand from Lynn, mentioned in the recipes of 1778-9, is still used by the British glass industry on account of its high purity. Another valuable source of sand was Alum Bay in the Isle of Wight.

The chemical revolution

It was not until 1736 that H. L. Duhamel-Dumonceau proved that common salt was a compound of the base of soda and 'spirit of salt' (hydrochloric acid) and in 1755 Joseph Black made clear the relationship between mild and caustic alkalis, showing that soda and potash were compounds of carbon dioxide with caustic soda and caustic potash respectively, thus paving the way for the manufacture of sodium carbonate from brine.

The next developments followed advances in the field of chemical analysis and the extensions in the range of known chemical substances. As we mentioned earlier, reliable methods for the analysis of minerals were developed by M. H. Klaproth at the end of the eighteenth and during the early years of the nineteenth centuries. He also discovered new elements such as uranium which could be used for the colouring of glasses. It now became clear that the old names such as crocus martis, tartar and ferretto were both confusing and inexact. Following Lavoisier's work on the naming of a compound according to the results of its analysis, a systematic nomenclature was established in 1787 in which symbols were given to various compounds. In 1807 Dalton's atomic theory was published. It described how identical atoms of an element combined with atoms of other elements in simple numerical proportions. But perhaps the greatest impact on glass technology was made by J. J. Berzelius who, from 1808 onwards, laid the foundations of quantitative chemical analysis and showed how the equivalents and atomic weights of elements could be used to make quantitative chemical predictions.

Chemical formulae and chemical equations were established which were immediately applicable to the glass industry.

The influence of chemical theory is shown in books and articles written about glass during the early nineteenth century. For example, Samuel Parkes says in his *Chemical Catechism* of 1826 that:

> The art . . . of making glass is . . . entirely chemical, consisting in the fusion of siliceous earth with alkali and the oxides of lead. (The manufacturer) will be enabled on chemical principles to ascertain the exact quantity necessary for any fixed portion of silica.

There was no longer the confusion in terminology which existed in many earlier texts and recipes. Manganese dioxide provides an example of this; it had been referred to as 'magnees', and was consequently confused with magnesia; but Humphrey Davy had demonstrated that the two were distinct when he showed that magnesia was a metallic oxide with a base of magnesium.

The chemical nature of soda and potash was by now well known by the glassmakers and some of the functions of lime in the glass were understood as an account of 1832 shows:

> Lime in the form of chalk is useful as a very cheap flux. It is also beneficial in facilitating the operations of the workman in fashioning glass, and it has the property of diminishing its liability to crack on exposure to sudden and great variations in temperature . . . Excessive lime (causes) the rapid destruction of the pots and renders the glass cloudy (when present above) six per cent.

In 1830 a technological article on the composition of soda-lime-silica glass was published by the French chemist, Dumas, in which he pointed out that the glass became more resistant to moisture attack as the composition approached the proportions of one part of sodium or potassium oxide to one part of calcium oxide, to six parts of silica. Modern soda-lime-silica based glass contains (by weight) seventy-two per cent silica, fifteen per cent sodium oxide, ten per cent calcium oxide, two per cent alumina and one per cent miscellaneous oxides. Such a glass is the optimum with respect to cost, durability and ease of manufacture, and contains approximately six molecular parts of silica to one of sodium oxide, and one of calcium oxide and magnesium oxide; the proportions found by Dumas.

The Bohemian glassmakers, on the other hand, had been using calcium carbonate in their 'fine Bohemian crystal' since the seventeenth century, and nineteenth-century recipes for crystal glass and 'very white window glass' seem to indicate that they were able to make a stable glass containing lime for luxury purposes. The following recipe for Bohemian crystal is given in a French mineralogical journal of 1843:

Pulverized quartz	100 parts
Calcined potash, first quality	55 parts
Carbonate of lime	8 parts
Arsenious acid (As_2O_3)	a trace
Peroxide of manganese (MnO_2)	¾ parts
Old glasses, broken and picked	50 parts

Apsley Pellatt, in his *Curiosities of Glassmaking* of 1849, gives the following recipe for flint glass which may be compared with the eighteenth-century recipe given earlier on page 63:

Carbonate of potash	1 cwt
Red lead or litharge	2 cwt
Sand washed and burnt	3 cwt
Saltpetre	14 to 28 lb
Oxide of manganese	4 to 12 oz

Although the textbooks of the period show the influence of the new knowledge, the private recipe books still used some of the older terminology; the following recipe for flint glass was found in a manuscript notebook written at about the same time as Pellatt's book:

2lb of flint batch	
Sand	16 oz
Lead	11 oz
Ashes	5½ oz
Nitre	½ oz

This notebook also contains recipes for many kinds of coloured glass, which was then becoming very popular. The use of chromium, uranium and nickel, all of which had been discovered during the latter part of the eighteenth century, is mentioned. Opaque glass was also popular and a variety of opacifying agents were used; tin oxide, calcined bones, oxide of antimony and oxide of arsenic. During the second half of the century, fluorides in the forms of fluorspar and cryolite, CaF_2 and Na_3AlF_6, were introduced.

The cheaper forms of glass, such as those used for bottles, were still made from very impure materials, the aim being to reduce the cost as much as possible. A recipe for black bottle glass quoted in a book of 1854 is given below:

Sand	3 barrows
Lime	4 barrows
Red clay	4 barrows
Rock salt	60 lb
Soap waste	28 lb

As the century drew to a close there was an increasing interest in glasses with special properties for specific purposes. This was particularly important in the rapidly developing field of optical instruments, the notable example of which is the work of Abbe and Schott at Jena, described in the next chapter. This interest in the properties of glass can be traced back to the eighteenth century when investigations started on such properties as strength, thermal expansion and chemical resistance. But these investigations were not pursued in the methodical and scientific manner which, together with co-operation between glass manufacturers, scientists and instrument makers, led to the supremacy of the German optical industry and ultimately to the growth of a scientifically based glass industry in the United States and Europe.

The following composition for a dense barium silicate crown glass is a later example of the type first produced by Abbe and Schott and shows the great changes which had taken place in the concept of batch compositions before the end of the nineteenth century. The quantities are given in mole per cent:

SiO_2	57.5
B_2O_3	6.0
Al_2O_3	4.5
BaO (+ trace of SrO)	25.0
PbO	0.5
ZnO	6.0
$ZrO_2 + TiO_2$	0.5
$Sb_2O_3 + As_2O_3$	0.3

Bibliography

1. *Analyses of ancient glasses, 1790-1957,* E. R. Caley, The Corning Museum of Glass monographs, Vol. 1, 1962.
2. *Studies in ancient glasses and glassmaking processes Part V: raw materials and melting processes,* W. E. S. Turner, *J. Soc. Glass Technol.,* 1956, **40**, 277.
3. *Ancient glass and glass-making,* W. E. S. Turner, *Proc. chem. Soc.,* 1961, 93.
4. *A dictionary of Assyrian chemistry and geology,* R. Campbell Thompson, O.U.P., 1936.
5. *The Rothschild Lycurgus Cup: an analytical investigation,* R. C. Chirnside, VIIth International Congress on Glass, Brussels, 1965.
6. *The Chemistry of the Lycurgus Cup,* R. H. Brill, VIIth International Congress on Glass, Brussels, 1965.
7. *The scientific investigation of ancient glasses,* R. H. Brill, VIIIth International Congress on Glass, London, 1968.
8. *Further historical studies based on X-ray diffraction methods of the reagents employed in making opal and opaque glasses,* W. E. S. Turner and H. P. Rooksby, Jahrbuch des Romisch-Germanischen Zentralmuseums Mainz VIIIth annual, 1961.
9. *The Art of Glass,* A. Neri, translated into English by C. Merrett, 1st edition, London, 1662.
10. *The Art of Glass,* H. de Blancourt, translated into English, London, 1699.
11. *The Laboratory, or School of Arts,* translated from the German, London, 1739.
12. *Bradshaw's Recipes,* a handwritten book of glass recipes, 1778-9.
13. *Glass,* Encyclopaedia Britannica, 3rd edition, 1797, vol. 7, 763-83.

14. *Memoir on the origin, progress, and improvement of glass manufactures,* A. Pellatt, B. J. Holdsworth, London, 1821.
15. *The chemical catechism,* 12th edition, S. Parkes, Baldwin, Cradock, and Joy, London, 1826.
16. *Treatise on the origin, progressive improvement and present state, of the manufacture, of porcelain and glass,* G. R. Porter, Longmans, London, 1829.
17. *Crown glass cutter and glaziers manual,* W. Cooper, Oliver and Boyd, Edinburgh, 1835.
18. *Curiosities of glass making,* A. Pellatt, D. Bogue, London, 1849.
19. *The useful arts and manufactures of Great Britain, part 1,* S.P.C.K., London, 1852.
20. *Treatise on the art of glass making,* 2nd edition, W. Gillinder, Birmingham, 1854.
21. *Nouveau manuel complet de Verrier,* vol. 1, J. Fontenelle and F. Malepeyre Roret, Paris, 1900.

4

Optical glass

The early history of lenses

It was not until the last quarter of the nineteenth century that glass technology became sufficiently advanced to make possible the development of a variety of satisfactory optical glasses, the introduction of which had important effects upon science and industry. The need for special glasses could not, however, be recognized until there was a certain degree of understanding of the use and design of lenses, an understanding which had been sought for over two thousand years.

Objects with the shape of lenses made from rock crystal and dating from 1600-1200 BC have been found in Crete, but they may have been used as ornaments. The Greeks knew of the power of lenses to concentrate the rays of the sun. In the play *Clouds* by Aristophanes (434 BC), a character who had been served with a summons written upon a wax tablet says that it could be melted by a burning glass. The Greeks also recognized that glass spheres filled with water could act as magnifying glasses and theories of reflection and refraction were worked out which partially explained these phenomena. Ptolemy, the Greek astronomer famous for his theories on the solar system, wrote a work on optics in the second century AD which included values of angles of incidence and refraction; but he had no effort to relate the two quantities by a general law.

Greek scientific knowledge later passed to the Arabic world where it was developed by many scholars. Notable in the field of optics was Ibn al Haitham, or Alhazen (965-1039 AD), who made the first mention of a magnifying lens in the form of part of a glass sphere. Alhazen described his experimental work in several books: his study of the rainbow was translated into Latin in about 1170 and his *Optics* in 1269. His work influenced several European scholars, notably Robert Grosseteste, Bishop of Lincoln (*c.* 1175-1253), and his pupil, Roger Bacon (1214-94). Grosseteste explained the colours of the rainbow by a theory of refraction which persisted until the sixteenth century. He proposed the use of a lens for magnification and bringing closer distant objects, and Bacon experimented with plano-convex lenses for spectacles.

The increasing interest in optics soon found practical expression in Italy around 1280 with the invention of spectacles with convex lenses for the correction of near sight. A Florentine script of 1289 mentions 'those glasses they call spectacles lately invented to the great advantage of poor old men when their sight grows weak'. Concave lenses, first mentioned by Nicholas of

Cusa in about 1450, did not come into general use until the mid sixteenth century. The lenses of the period were not designed scientifically and choosing a pair of spectacles seems to have been a process of trial and error on the part of the buyer. Venetian guild codes for glassmakers of 1300 mention 'little discs for the eyes' and in 1301 'eye-glasses for reading'. The spectacle-makers were able to produce finished lenses without much difficulty using grinding and polishing techniques evolved for finishing the surfaces of jewels.

Glass for telescopes and microscopes

According to a manuscript of 1634 a telescope was made in about 1590 in Italy which was at the time the main centre for glassworking and optical studies. It was used as a model by a Dutch spectacle-maker, Janssen, who constructed his first telescope in 1604. There is no doubt that telescopes were used by the Dutch for military purposes about this time but three centuries had elapsed between the invention of the lens and the invention of the telescope. Concave lenses for the correction of short sight were not widely used until the mid sixteenth century. The first simple telescopes were composed of a tube with an ordinary convex spectacle lens at the front as objective and a concave lens as the eyepiece. They were about one foot long, with a magnification of about three times.

In 1609 Galileo learned of the Dutch invention and he soon built his own telescopes, finally achieving one with a power of x32, with which he was able to observe Saturn's rings, the satellites of Jupiter and the mountains on the moon; but the field of view of his telescope was very limited.

A similar lens combination, with lenses of suitable focal length, could also be used as a form of compound microscope. In 1610 Johannes Kepler first gave a description of the combination of two convex lenses, which gave an inverted image, in a telescopic arrangement with a very substantially enlarged field of view.

Intrinsic defects in lenses

If a lens is used to focus light from a distant object, rays passing through the edges of the lens come to a different focus from those passing through the lens centre. This gives an imperfect image (Figure 16a). The defect, known as spherical aberration, can be cured by using a lens with a non-spherical surface; such lenses have been made since the eighteenth century, but it is still difficult to make them both accurately and cheaply. During grinding, the lens blanks are mounted on a curved surface and the assembly is moved over a similar iron surface with abrasive slurry in between until all restrictions have been broken away; this can only be done when both surfaces are spherical, so that one can slide over the other freely in any direction. The iron support for the lens blank is produced in the same way, thus spherical surfaces are relatively simple to produce.

A second type of aberration is known as chromatic aberration (Figure 16b). The refractive index of glass is different for different wavelengths of light. This causes the different wavelengths to come to slightly different focal points, resulting in a coloured fringe to the image. In nature the variation of refractive index of water with wavelength gives us the beauty of the rainbow; in lenses the variation causes chromatic aberration. The problem can be greatly reduced by using lenses of large focal length and the astronomers of the seventeenth century therefore built very long telescopes: instruments up to 210 feet long were constructed in great variety. Figure 17 shows a novel attempt to solve the problem.

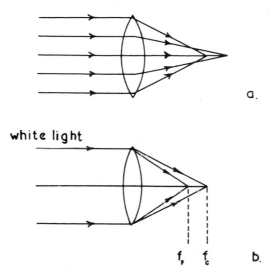

white light

f$_F$ f$_c$ b.

Fig. 16. (a) Spherical aberration in a lens—rays of light passing through the margins of the lens do not come to the same focus as rays passing through the centre of the lens. (b) Chromatic aberration in a lens—light of different wavelengths comes to a focus at different points on the axis of the lens. The foci for red light (f$_c$) and blue light (f$_F$) are shown.

The cause of chromatic aberration was first ascribed correctly to the variation of refractive index with wavelength by Newton in 1671. When a ray of white light, composed of all the spectral colours, passes through a lens each wavelength is bent, or deviated, by a different amount and again when the ray leaves the lens. The angle of deviation is defined as the angle between the incident ray and an emergent ray which is located near the centre of the visible spectrum, such as yellow light with a wavelength of 5893 Å (spectral line D). The angle of dispersion is defined as the angle between the emergent extreme rays of the spectrum, usually taken as red light with a wavelength of 6563 Å (spectral line C), and blue light with a wavelength of 4861 Å

The Aerial Telescope.

Fig. 17. Early telescopes gave imperfect images owing to defects in the lenses. In order to lessen these effects the telescope makers produced very long telescopes which had to be supported by extensive scaffolding or, as in this picture, suspended from a tall tower or building.

(spectral line F). Dispersion and deviation can be discussed in terms of the refractive indices, N_D, N_C and N_F respectively, for the three wavelengths given above, the refractive index being defined as the ratio of the sine of the angle of incidence to the sine of the angle of refraction. Then the dispersive power, or relative dispersion, is defined as:

$$\frac{N_F - N_C}{N_D - 1} = V$$

V measures the amount of dispersion relative to the amount of deviation. In optical catalogues the value usually quoted is the reciprocal of V, $(1/V)$, which is known as the constringence. Modern optical glasses have values of constringence ranging from sixty-five (low dispersion) to twenty-one (high dispersion).

Chromatic aberration is corrected by combining a converging and a diverging lens so chosen that the red and blue images formed by the system coincide on the optical axis. Although this does not completely remove the aberration, because the system is only corrected for two specified wavelengths, by careful choice of glasses and design of lenses the aberration can be greatly reduced. The conditions for achromatic correction are that the focal lengths of the lenses, f_1 and f_2, and their relative dispersions, V_1 and V_2, obey the relation:

$$\frac{f_1}{f_2} = \frac{V_2}{V_1}$$

The glasses available to Newton did not differ much in relative dispersion and unfortunately his observations led him to believe that all glasses have the same relative dispersion, thus making it impossible to correct chromatic aberration in a lens system. It was not until seventy-five years later that the first achromatic doublet was made.

Having decided that chromatic aberration could not be corrected, Newton turned his attention to reflecting telescopes which do not suffer from this fault, and as a result of his opinions refracting telescopes soon lost their popularity. Newton constructed his telescopic system using a spherical mirror, which although not suffering from chromatic aberration was still subject to spherical aberration (Figure 18a). As with lenses, the spherical shape was by far the easiest shape to manufacture but a parabolic mirror removes spherical aberration (Figure 18b) and the first parabolic reflecting telescope was made in 1723 by John Hadley. A parabolic mirror has the property of bringing to one focus on its axis all rays of light which travel parallel to the axis prior to reflection. It is thus much more satisfactory for astronomical telescopes than a spherical mirror.

The difficulties encountered in making parabolic mirrors produced a movement in favour of the refracting telescope, which grew even longer. Increased efforts were made to correct chromatic aberration and in the 1730s

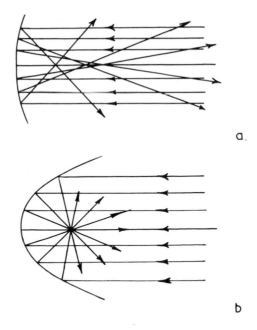

a.

b

Fig. 18. (a) Spherical mirror—each ray of light passes through a different focal point and the image is blurred. (b) Parabolic mirror—all rays of light travelling to the mirror from a distant object are reflected through the same focal point and the image is sharp.

Chester Moore Hall invented the achromatic object lens by combining a converging crown lens and a diverging flint lens. He had carried out his experiments in the belief that correction of chromatic aberration must be possible, in spite of Newton's views, because the lens system of the eye does not appear to produce a falsely coloured image; in fact, there is chromatic aberration in the eye but only to a tolerable degree.

The achromatic doublet was made possible when flint glass, Ravenscroft's 'glass of lead', became available in England at the end of the seventeenth century. Eighteenth-century flint glass had a high lead content and a correspondingly high density; consequently the refractive index of flint glass was much greater than that of crown glass. The dispersion was also much larger, as required for the production of an achromatic doublet.

A patent for an achromatic lens, consisting of a cemented combination of convex crown and concave flint lenses, was granted to the optician and instrument-maker John Dolland in 1758. He may have learnt of Chester Moore Hall's invention of the doublet from a grinder who had made the lenses for Moore Hall; Dolland's lenses were made commercially but were of poor quality. Dolland's son enforced the patent with the result that the achromatic lens remained inferior through lack of competing products and was not in general use in microscopes for another seventy years.

Until the end of the eighteenth century the glass used for lenses was obtained as a by-product of other sections of the glass industry. The lenses were small and generally of poor optical quality, often containing inhomogeneities in the form of bubbles and unmelted batch materials, and tinted owing to impurities in the glass. Lenses were cut from thick crown or plate glass, hence the term 'crown' glass originally used for optical glass of a soda-lime-silica composition.

Homogenization of optical glass by stirring

A major advance in optical glass manufacture was made at the end of the eighteenth century by the Swiss worker, Pierre Louis Guinand. He was originally a woodworker who made clock cases and later developed his skills in metal-working and made bells; his interest in glass for telescope lenses dated from about 1768, and he made many series of melts, but with little success. The revolutionary idea of stirring the glass to improve its homogeneity probably occurred to him when he was engaged on the casting of bells, because the molten metal used for the bells had to be stirred prior to casting for exactly the same reason. Another story suggests that he was inspired by the stirring of 'fondue', melted cheese mixed with butter or eggs to form a smooth savoury mixture.

Guinand first used a stirrer in the summer of 1798, after thirty years of experiments. This stirrer was of a mushroom shape and was not too successful, but by 1805 Guinand had devised a hollow cylindrical rod of porous burnt fireclay which was later widely adopted. The same principle is still used for the manufacture of optical glass although it is applied under far more strictly controlled conditions, stirring being the only effective means of making glass homogeneous. The time taken for homogenization by the inter-diffusion of the different components of the glass is so long that it is necessary to stretch out the inhomogeneous regions by a stirring action until their dimensions are very small: they then have to move only very small distances for inter-diffusion, and hence homogenization, to occur. The need for stirring can easily be seen by watching sugar dissolve in a glass of water. In glass all the diffusive processes involved would take about ten thousand times longer.

The problem of homogenization applies especially to flint glass, as the lead oxide with its relatively high density tends to be distributed unevenly through the glass. Stirring enables much larger quantities of heavier materials to be added to the batch, making it possible to extend the composition range; prolonged stirring also helps to free the glass from gas bubbles formed during melting.

Guinand moved to Benediktbeurn in Bavaria to manage the glassworks of Utzschneider and from 1805 worked closely with the young scientist Joseph von Fraunhofer. They improved the stirring process and extended the range of dense flint and crown glasses but glass was not produced on a large scale. Guinand returned to Switzerland in 1814 and died there in 1824. The secret of the stirring process was passed on to his son Henri, who founded an optical glass factory in Paris in 1832. By now other countries were becoming very interested in making optical glass and were anxious to obtain details about the stirring process, but it was not until 1848 that commerical optical glass manufacture was started in England by Chance Brothers of Birmingham.

The work of Fraunhofer, Faraday and Harcourt on the melting of optical glass

Prior to the introduction of optical glass manufacture into England, Michael Faraday and John Herschel were asked by the Royal Society to study the preparation of optical glasses after reports had been received concerning the progress that Fraunhofer had made in his investigations of the dependence of refractive index upon the conditions of manufacture. A committee was appointed in 1824 by the Royal Society with the task of improving glass for optical purposes, and Faraday started his experiments at the Falcon Glass Works, Blackfriars. The government agreed to waive excise duties for the experiments and to bear the costs of furnaces, materials and labour as long as the work showed promise of practical success. The work was moved to the Royal Institution in 1827.

Faraday's work was at first concerned with both flint and crown glass, but from 1828 onwards he concentrated on flint glass in which more veins and striae were formed by the solution of the pot by hot glass. In an attempt to solve these problems Faraday used batches containing silica, boracic acid and lead oxide which he melted in a platinum container and stirred with a platinum stirrer. He paid great attention to the purity of his raw materials, which led him to discover the effect of manganese in producing a deep purple colour in his glass, an effect which was far more marked than in normal flint glass. He also noted that iron caused a very strong colour in his glasses and stressed in his reports the importance of keeping optical glass free from metallic contamination during preparation.

Faraday succeeded in preparing a wide range of glasses with high specific gravities, even as high as 6.4, about twice that of ordinary flint glass made at the time. This heavy glass was a mixture of lead and boric oxides containing from sixty to eighty-two per cent lead oxide. He was able to use his glass to make object lenses of high optical quality.

Unfortunately his meticulous work had little practical effect at the time, and it ceased in 1830 when the industry found that good optical glass could be imported. His work on lead borosilicate glass was not wasted. On the 13th September 1845 he observed an interaction between light and a magnetic field, now known as the Faraday effect, which inspired Maxwell to produce his electromagnetic theory of light, an outstanding achievement in the classical physics of the nineteenth century. To observe the Faraday effect it is necessary to produce light polarized in one plane, a process familiar nowadays in polaroid spectacles but achieved at the time by reflection at a critical angle or by passing the light through a tourmaline crystal which, like polaroid, only transmits the components of light vibrating in one direction. Such polarized light was then passed through a block of the heavy flint glass in a magnetic field and as a result the plane of polarization was rotated. This effect has been used in recent times in the technology of lasers.

From 1834 the Reverend William Vernon Harcourt made a long series of experiments on the composition and melting of optical glass. His melts were made on a very small scale in platinum crucibles, the components being chosen from some thirty different oxides; but he concentrated particularly on phosphate glasses containing oxides of potassium, sodium, lithium, aluminium, calcium, strontium, barium, titanium, molybdenum and tungsten. After twenty-five years of work, carried out from 1862 in collaboration with Sir George Stokes, Harcourt was able to make phosphate glasses containing titanium which could be used in lens systems to give improved achromatic qualities. His melts were made in a small crucible rotated by clockwork and revolved within a spiral burner using hydrogen as the fuel, but he did not stir the glass, which probably accounts for his consistent failure to obtain glasses with reproducible optical properties on which accurate optical measurements could be made. In 1874 Stokes was able to show a very small triple objective lens to the British Association which was free from secondary spectrum effects (see page 87). Unfortunately the quality of Harcourt's glasses was poor and his work failed to make an impression upon the manufacturers of optical glass, although a member of the firm of Chance Brothers, John Hopkinson, aided Stokes in his experiments.

The introduction of optical glass manufacture into England

During this period of growing interest in improved optical glass, many attempts were made to learn details of Guinand's secret process. The glass technologist Georges Bontemps, who owned a glassworks at Choisy-le-Roi, was anxious to win a prize for glass offered by the French Academy of Science. He was introduced to Henri Guinand by the optician Lerebours and the three made an agreement in 1827 for the purchase and development of the secret by Bontemps. Unfortunately Guinand appears to have been less well informed on the subject of stirring than his father and the agreement was terminated after unsuccessful experiments at Choisy-le-Roi. Bontemps continued to work on his own and in 1828 was able to present to the Academy of Science satisfactory discs of optical glass twelve inches in diameter.

The work of Bontemps and Lucas Chance of Chance Brothers, Birmingham, in the development of flat glass will be discussed in Chapter 6.

Their association in this field started in 1832 and in 1837 the same friend of both parties who had helped to introduce sheet glass manufacture to Britain, A. Claudet, opened talks with Bontemps on optical glass. The factory at Choisy-le-Roi had by now been manufacturing optical glass regularly for ten years, and discs of both flint and crown optical glass of considerable size were now made. Negotiations for the manufacture of optical glass at the Spon Lane works of Chance Brothers at Smethwick started in May 1837. It was agreed that Bontemps should be paid for his knowledge a sum of 3000 francs, the amount that he had originally given to Guinand for the same information; this sum was to be paid when Chance Brothers had made the same amount as profit. A patent for Guinand's process was taken out by Lucas Chance in March 1838.

For several reasons the work progressed slowly. Development of the new patent plate process (see Chapter 6) took up much of the time available for trials, and the transfer of information between Chance Brothers and Bontemps in France was unsatisfactory. In June 1840 Lucas Chance invited Bontemps to England to perfect the process and Bontemps in accepting the invitation proposed sending in advance one of his expert workmen. William Chance objected strongly to the agreement, believing that the optical work was far less important than other schemes in progress and also that everything should be done within the organization of the firm. His views prevailed over those of his brother and optical glass manufacture was postponed for eight years.

In 1848 Bontemps was forced to leave France by events connected with the Revolution which deposed King Louis XVIII, and he entered into an agreement with Chance Brothers to devote all his services to them in many of their enterprises. When he arrived, Bontemps set to work immediately producing soft crown and light flint for camera lenses, and hard crown and dense flint for telescopes. Hard crown had a similar composition to ordinary crown glass, but potash was substituted for soda as the alkali; for soft crown the proportions of both lime and red lead to sand were 9.66 parts to 100 parts of sand. The old notebooks of Chance Brothers give this information, but the alkali contents of Bontemps' subsequent glasses are not given. The first dense flint contained equal quantities of sand and red lead, but Bontemps gradually increased the lead content until in 1867 he introduced double extra dense flint in which nine parts of red lead were used to five parts of sand. Following his work, six types of flint glass, containing varying proportions of lead were manufactured by Chance Brothers, and with some modifications in the amount of alkali, remained as their standards until the 1870s when many new optical glasses were introduced by Schott at Jena.

The optical side of the business soon became very prosperous, and Bontemps made several visits to the Continent, returning with substantial export orders. At home the new glasses won approval as Bontemps reported:

The glass is approved and the customers expressed their satisfaction not to be obliged to bring their materials from abroad ... (a most competent optician) says that our flint glass is even superior to the Swiss flint, not being altered so easily by the atmosphere. ... As for the British flint plate, they confess that our light flint is superior by far.

Chance Brothers were able to make an impressive display of optical glass at the Great Exhibition of 1851. The outstanding exhibit was a 29-inch disc of dense flint, $2\frac{1}{4}$ inches thick and weighing about 200 lb. It was almost completely free from veins, striae and bubbles and Chance Brothers showed that a perfect lens of up to 25 inches in diameter could be formed from it. The disc gained a Council Medal, the judges considering that 'as the great object-glasses of Pulkowa and of New Cambridge in the United States were no greater than 16 inches in diameter, the present disc was an outstanding achievement'.

The development of the optical instrument industry

The success of Chance Brothers in the manufacture of optical glass reflected the increasing demands made upon glass by the optical instrument industry. By the middle of the nineteenth century high quality optical glass was being made in England, and the work of Newton, Moore Hall, Dolland, Faraday and Harcourt would suggest that England was well placed to take the lead in developing and supplying new glasses and new optical systems. In fact, by the 1850s, England had lost her prominent position of the eighteenth century and her place was being taken by Germany.

The reasons for this change are complex, but a major contributory factor was the willingness of the German scientists, glass manufacturers, opticians and instrument makers to co-operate and exchange information. When a new range of optical glasses had been developed by Abbe and Schott, the German government provided constant support to the growing industry, especially in the field of education. During the last quarter of the nineteenth century professional schools were founded in many towns. Attached to the schools were scientific laboratories and workshops for instrument manufacture, thus ensuring close contact between students, scientists and mechanics. In England, on the other hand, no university training in technical optics was available until, in 1917, Imperial College, London, instituted a programme on applied optics.

The rise of the German instrument industry may be said to have begun with Fraunhofer's work on improving optical glasses at Benediktbeurn (1805-25). Although he was unable to devise a completely reliable commercial process for the production of optical glass he is a typical example of a scientist who worked closely with glass manufacturers and instrument

Fig. 19. The Zeiss workshop, 1864. The master craftsman (extreme right) is teaching his apprentices to grind, polish and test lenses. The tools hang in rows at the back of the workshop and the workers operate the machines by means of treadles.

makers in nineteenth-century Germany. In 1807 he joined the Mechanico-Optical Institute which had been founded in 1804 by the glass manufacturer Utzschneider, together with Reichenbach and Liebherr. The latter two owned an optical instrument workshop in Munich. This type of co-operation was noticeably lacking in England, where the first 'Optical Society' was not founded until 1889. The British Scientific Instrument Association was formed in 1918.

Whilst Fraunhofer and his associates were laying the foundations for the successful growth of the German optical industry, Britain was experiencing great difficulties in maintaining her position in the field. A major factor affecting development, especially of astronomical telescopes which required large lenses, was the heavy tax imposed by the government on flint glass. The repercussions of the general taxes on glass during the eighteenth and nineteenth centuries have been discussed in Chapter 2, and the optical industry was one amongst many to suffer from these repressive measures. Glass taxes were lifted in 1845 but the government policy can hardly be said to have changed to one of encouragement and in the 1850s Britain had been outstripped by France and Germany in the manufacture of telescopes, although, until the closing years of the nineteenth century, the British microscope objectives were judged to be the best made.

The quality of British glass was excellent but the industry was, in general, content to produce the same types of glass as long as there was no demand from the customers for anything different. The Chance standard glasses of the 1850s and 60s appear to have fulfilled all British requirements and it was not until the late 1880s that their sales declined as both British and foreign instrument makers started to use the wide range of improved optical glasses from Germany. For over thirty years no demands were made in Britain for new types of glass by the opticians, who were still, for the most part, craftsmen in the eighteenth-century tradition.

Abbe, Schott and Zeiss

The German instrument maker Carl Zeiss opened an optical workshop in 1846 in Jena, and from the first he had close associations with the University of Jena. The workshop was mainly concerned with the building of microscopes, at the same time collaborating with several scientists, especially those attached to Jena University. Zeiss was particularly influenced in his early work by the Professor of Botany, Matthias Jacob Schleiden, who encouraged him to build microscopes of high quality which successfully competed with those of other countries, especially France, England and America. Figure 19 shows members of his workshop in 1864, employed in forming and testing lenses.

By the 1860s, Zeiss had gathered a vast amount of experience in the manufacture of optical instruments and he became the mechanic at the

University, where he met the scientist Ernst Abbe, the Professor of Physics, who eventually joined him in the development of optical glasses. Abbe finally went into partnership with Zeiss in 1875 but he retained close connections with the University as a teacher until his death in 1905.

At that time lens-making was a craft and advances were made by trial and error. Zeiss believed that optical instrumentation should be placed on a scientific and mathematical basis and persuaded Abbe to investigate the physics of image formation by lenses. Abbe made many fundamental discoveries in physical and geometrical optics which were applied to the manufacture of microscopes. He also developed a series of new instruments for the inspection of optical and mechanical components and after a long period of experiment the workshop had great success in the manufacture of scientifically designed microscopes and drew ahead of other optical instrument firms.

However, Abbe realized that the full theoretical advantages could not be obtained because the optical glasses available had such a limited range of physical properties. He wrote in 1876:

> The future of the microscope as regards further improvement in its dioptric qualities seems to be chiefly in the hands of the glassmaker. The especial desiderata are a distribution of colour dispersion more favourable to the removal of the secondary spectrum and a greater variability in the relation between dispersion and mean index. . . . The limitations observed in the connection between refraction and dispersion in existing glasses must not be regarded as a natural necessity; for among both natural and artificial products, there are plenty of transparent substances which are known to have widely different purposes as regards refraction and dispersion. . . . Stokes in England . . . gave useful hints as to the specific effects of certain bases and acids (in the production of optical glasses) on the refraction of light. The uniformity shown by existing glasses in their optical qualities is probably chiefly due to the very limited number of materials hitherto used in their manufacture. Beyond silicic acid, alkali, lime and lead, scarcely any substances have been tried except perhaps alumina and thallium.

Abbe went on to describe how progress was impeded by the monopolistic nature of the manufacture of optical glass and consequent lack of competition. Advances were confined to traditional lines and change could only be brought about, firstly by public financial support for costly experiments, and secondly by scientific help from learned societies. His plans for future work were also stated:

> When this narrow groove (of composition) is left, and a methodical study, on an extended scale, is made of the optical qualities of chemical elements in combination, we may anticipate with some confidence a greater variety in the products.

The remarks of Abbe were read with great interest by Otto Schott, a German chemist who had investigated chemical reactions involved in glass melting. Schott, who was a member of an established glassmaking family, sent some samples of lithium glasses which he had prepared to Abbe, hoping that they would be of use to him in his investigations. In 1881 they began joint research into the problem.

A new range of optical glasses

The first glasses were melted by Schott at Witten on a very small scale, his aim being to study the influence on optical properties of as many new glass components as possible. The glasses were then examined by Abbe in Jena. By the end of 1881 very promising results had been obtained and it was necessary to carry out experiments on a larger scale with a view to commercial production. In 1882, Schott went to Jena where a special laboratory was set up complete with all the necessary apparatus such as gas furnaces and electric motor blowers which had first become available in 1873. Here they were able to make melts of up to ten kilograms, and at a later stage, twenty-five kilograms. They also had the services of an analytical chemist and a skilled glass worker, and for nearly two years, until the end of 1883, they devoted their attention to solving problems which arose from the requirements of practical optics. In a trade catalogue of 1886 they described those requirements as follows:

> One was the problem of producing crown and flint pairs with as nearly as possible proportional dispersion throughout the different sections of the spectrum, in order to render possible a higher degree of achromatism than the glasses hitherto in use permitted, and thus abolish or diminish the strong secondary chromatic aberration which silicate glasses by reason of the different distributions of dispersion in crown and flint, are never able to remove in achromatic combinations.
> The second problem ... which ... has not hitherto been so generally recognized, was the attainment of greater diversity in both the ... mean index and mean dispersion. ... The systematic use of a larger number of chemical elements in the composition of glass has rendered such gradations possible (and thus) the choice between available glasses, instead of being substantially of a linear character as heretofore, has become two-dimensional.

At the start of Abbe's and Schott's experiments, the optical effects of only five glassmaking oxides were well known, these oxides being silica, potash, soda, lead oxide and lime. As there was a tradition at the time that boric oxide and phosphoric oxide, both known glassmaking acid oxides, gave readily tarnished or not very durable glasses, Abbe and Schott began by testing the optical properties of phosphoric and boric oxide in combination with a wide range of metallic oxides. To the elements silicon, potassium, sodium, lead, calcium and oxygen, they gradually added twenty-eight other elements in quantities of at least ten per cent; in doing so they found glasses in which the relation between refractive index and dispersion was very different from the old glasses. However, addition of most of these elements did not enable pairs of glasses to be produced with proportional dispersion. Further experiments showed that by adding a large percentage of boric acid to the flint glass batch and by using fluorine as a crown glass component, the partial dispersions in various regions of the visible spectrum could be altered and pairs of glasses obtained with greatly improved achromatism.

Unfortunately, addition of fluorine to the glass batch caused it to give off pungent fumes during melting and the glasses were not very homogeneous. But by the autumn of 1883 both problems had been effectively solved and commercial production of successful new optical glasses such as the borosilicate crowns, barium crowns, barium flints and borates and phosphate glasses soon followed. The government was most impressed by the work and gave large grants to the laboratory which in 1884 became the Jena Glass Works of Schott und Genossen.

The success of the firm was spectacular. Its first price list of 1886 contained forty-four optical glasses of which nineteen were essentially new compositions. The first supplement of 1888 added twenty-four glasses, including eight new baryta light flints which were remarkable for their small dispersion compared with refractive index. They contained so little lead oxide that the usual light absorption shown by flint glasses was greatly reduced. New glasses were added to the list every few years and the effect on the manufacture of optical systems was so great that Germany, which had previously imported ninety per cent of her optical glass from England and France, started to export to these countries. Thus an industrial development which was accomplished in less than ten years virtually eliminated existing manufacturers and for about thirty years, until the outbreak of the First World War, Jena held an effective world monopoly in the manufacture of optical glass.

Improvements in the microscope

The effect of the new discoveries on optical instruments may be illustrated by Abbe's own work. Using the technical resources of the Zeiss works and the new glasses, he was now able to fulfil his plans for improving the microscope. Previous lens combinations of crown and flint glass had only been able to reduce chromatic aberration to a limited degree, as they matched only two particular wavelengths of the spectrum. The residual or secondary spectrum caused by the lack of common focus for other wavelengths was extensive. Figure 20 shows how the focal length varies with wavelength for a combination of the best available old English hard crown and dense flint (1) and for a doublet of phosphate crown with borate flint (2). The horizontal lines marked C, D, F and G′ refer to the wavelengths for red, yellow, blue and violet light respectively. For the crown and flint combination it can be seen that the characteristic curve intersects the C and F lines at the same value of focal length, *i.e.* the red and blue rays come to the same focus, and the lens system is corrected for chromatic aberration at these two wavelengths.

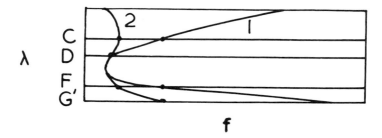

Fig. 20. Variation in focal length, f, with wavelength, λ, for a doublet for (1) old English hard crown and dense flint glass; (2) phosphate crown and borate flint. The borate-phosphate doublet suffers much less from chromatic aberration than the traditional crown-flint doublet and when used in microscopes gives higher magnification and improves image quality.

However, the variation of focal length with wavelength over the whole of the visible spectrum is considerable, resulting in a large amount of chromatic aberration, the secondary spectrum. The phosphate crown and borate flint combination, on the other hand, shows a much less rapid variation of focal length with wavelength and the focal lengths for red, yellow and blue light are nearly equal. Thus the combination is corrected for three wavelengths, and the residual spectrum, known as the tertiary spectrum, is much less extensive than the secondary spectrum of the traditional crown-flint doublet.

Abbe gave the name apochromatic to objectives formed from borate and phosphate glasses in which the secondary aberration had been removed. The critical amplification, *i.e.* the highest magnification that the lens can bear without loss of sharpness, was at least twelve to fifteen for the largest apertures, and considerably greater for medium and small apertures. The improvement in critical amplification for a smaller lens aperture arises from the fact that the wider the aperture through which the light is admitted to the lens system, the more difficult it is to form a good image. The values for the borate and phosphate objectives may be compared with the best results that Abbe could obtain for the old silicate glasses of from four to six. In the apochromatic lens it was also possible to correct spherical aberration for two colours, whilst in the silicate lens systems correction could be obtained only for one wavelength.

This improvement in performance meant that an apochromatic system could extend the range of magnification and at the same time improve the quality of the image. Abbe also developed many other systems for improving microscope performance such as an illuminator with a condenser lens, compensating eyepieces, and blood-counting and drawing attachments.

The new microscopes were soon employed in biological and bacteriological research which was just beginning at that time. Without them, indeed, such research would have been impossible because a great improvement in image size and quality was necessary in order to investigate the minute organisms. Prior to this time some of the most notable work had been done by Antony van Leeuwenhoek (1632-1723) who was both an instrument-maker and a gifted biologist. He ground his own lenses to give the best magnification in microscopes then available and made numerous discoveries using his own instruments. He described accurately blood corpuscles and observed the structure of muscle fibres, spermatozoa, skin, hair, teeth, and the eye. He also made important zoological investigations, including those on infusoria, rotifera, ants' eggs, pupae, cochineal, fleas, mussels and eels. This vast amount of work occupied him until his ninety-first year.

When the improved microscopes designed by Abbe became available it was possible to investigate even more minute structures. Dr Robert Koch, who is generally regarded as the founder of bacteriological technique, stated in 1904 that a large part of his success was due to the excellent microscopes of Abbe. His discoveries revolutionized the medicine of the day, for he discovered the cholera and tubercule bacilli, and showed that a bacillus was the cause of anthrax.

The First World War—optical glass as a strategic material

At the outbreak of the First World War supplies of German glass to France, the United States and Great Britain ceased. Great quantities of optical glasses were required for field glasses, cameras, gun-sights and other military instruments, and a critical situation arose. Although optical glass was being manufactured by Parra-Mantois in France, Bausch and Lomb in the United States and Chance Brothers in England, it was not available on either a large enough scale or in a sufficiently wide range of compositions.

Chance Brothers had, indeed, continued research throughout the period of the Jena work. In 1895 they started experiments on the production of some of the Jena glasses and obtained a satisfactory borosilicate crown glass. Their work on light flint and dense crown baryta glasses was not so successful, as the impurities in the barium materials coloured the glasses, which also contained a lot of small bubbles. Although these difficulties had been overcome by 1914 little progress towards commercial production had been made.

However, the situation quickly changed when the War Office discovered that suitable optical glass could be made in England. Funds were granted for extra buildings, machines and a research laboratory. Chance's output of optical glass rose from 2600 lb for the first half of 1914 to 92,000 lb for the first half of 1918, and the firm gave help to Russia in the establishment of its optical industry. Later, in the Second World War, they played a very important part in establishing optical glass production in Canada.

Many important technical problems were solved and a zinc crown, a dense barium crown and a light barium flint were commercially available late in 1914. New difficulties arose when essential supplies of potash salts from Stassfurt were cut off, but the British Cyanides Company were soon able to supply very pure, but extremely expensive, potassium carbonate. The Company also succeeded in purifying its own supplies of barium minerals to give the pure carbonate and nitrate.

The first topic investigated in the new research laboratory of Chance Brothers, which came into full operation in June 1917, was the opalescence and discoloration of fluor crown glasses; this resulted in the development of a glass having less dispersion than any of the Jena glasses. At the end of 1917 a sudden demand arose for a very large number of photographic lenses for the Air Board, requiring the production of several barium crown and flint glasses. As a result of intensive work on such problems, light, medium and dense baryta glasses and borosilicate crown glasses were being regularly produced by 1918 and many improvements had also been made in melting techniques, rapid measurement of refractive indices and testing of glasses.

In about 1917 annealing of optical glass was placed on a scientific basis by the work of F. Twyman who measured the variation of strain in glass as a function of time and used his results to determine suitable annealing schedules and temperatures. His work was followed by that of Adams and Williamson in the USA, who published an extensive mathematical study of the annealing process and also made detailed theoretical studies of homogenization by stirring.

The requirements of war brought about a close co-operation between the Geophysical Laboratory of the Carnegie Institution of Washington, engaged at that time in phase equilibrium studies on silicate systems, the National Bureau of Standards, working on glass and ceramics, and optical firms such as Bausch and Lomb. This co-operation resulted in many improvements in the manufacture of optical glass and by the end of the war the American industry had progressed to such an extent that from importing its entire supply in 1914, it was able in 1918 to supply Italy's requirements and so remove a considerable burden from France and Great Britain.

The investigations in the laboratories and works of the United States covered all aspects of production. G. W. Morey shortened the time required for melting from three days to one day, mechanical stirring was introduced in 1915 by Bausch and Lomb and later, vertical motion for the improved stirring of flint glass. The process of annealing was for the first time given a sound physical basis and polarized light was first used for the examination of the residual strain in glass. As glass was in such short supply methods were sought for producing it in quantity, which led to the rolling of optical glass into sheets. This process was satisfactory at that time for certain optical elements where fine striae in one plane only, as present in the rolled sheet, did not affect the optical performance. Soon many kinds of optical glass were being made by this method.

After the war many scientific papers appeared describing these investigations and adding greatly to the available knowledge of the relationships between physical properties and chemical composition. A new era in the accumulation of scientific knowledge and glassmaking had begun.

Rare earth and fluoride glasses

During the 1920s and 1930s new developments in optical glasses resulted in a great expansion in the types of glass available to the optical designers. Figure 21 is a graph of refractive index, n, plotted against constringence, ν; a particular glass can thus be represented by a single point on this plot. Prior to the work of Abbe and Schott, the range of available glasses was extremely limited; they were of the traditional crown or flint type and were characterized by a linear relation between dispersion and refractive index. These glasses are represented by the dotted area in the figure.

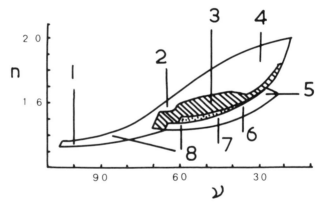

Fig. 21. A graph of refractive index, n, versus constringence, or reciprocal relative dispersion, ν, for a range of optical glasses. 1. fluoride glases; 2. fluoborate glasses; 3. ordinary optical glasses; 4. rare element borate glasses; 5. fluogermanate glasses; 6. fluosilicate glasses; 7. titanium glasses; 8. fluophosphate glasses. This wide range of glasses with independently varying dispersions and refractive indices enables modern designers to devise versatile optical systems.

The hatched area of Figure 21 shows how far the range had been extended by 1934, largely due to the work of Abbe and Schott. Glasses were now available which did not show a linear increase of dispersion with refractive index, and from these glasses, lenses could be made with a negligible secondary spectrum.

The plain area represents the glasses which have been introduced since 1934, and it can be seen that they form by the far the greater part of the modern range of optical glasses. Many of these glasses were produced as a result of the work of G. W. Morey and his colleagues in the United States. Morey was able to show that optical glasses need not be based on silica as the glass former (the main structural component of the glass). Instead he succeeded in using boric oxide alone as the glass former with additions of rare earth oxides such as lanthanum oxide which gave optical glasses with very high refractive indices but low dispersions.

Rare-earth borate glasses are extremely corrosive of refractory materials and melting therefore had to be carried out in plantinum or platinum-lined pots, a procedure adopted as far back as the 1820s by Faraday. The initial expense was justified because the pots could be used over and over again and waste glass recovered, an important consideration in view of the high cost of some of the batch materials. Production began in the United States just before the Second World War and the glasses were used to make extremely large and accurate photographic lenses.

The range of available dispersions and refractive indices was further extended by the development of fluosilicate flints, or super flints, which are glasses with dispersions greater than conventional flint glasses having the same refractive index. From Figure 21 it can be seen that an important portion of the range has been filled in by glasses such as the fluoborates, phosphates and germanates, and even types of all-fluoride glasses which use fluorine as the glass former instead of silicon, boron or phosphorus. All-fluoride glasses, with their very low dispersion and refractive indices, have an extended range of radiation transmission in both the infra-red and ultra-violet, making them very useful as components of optical systems. Thus, modern optical designers have available to them a very wide range of glasses with independently varying dispersions and refractive indices. Whilst this range enables them to design versatile optical systems, they actually require fewer glasses with slightly differing compositions than previously, as they are aided in their design work by the use of digital computers.

The melting of optical glass

The methods of production of optical glass changed very little until the Second World War although several improvements were made. Machine stirring replaced hand stirring after the First World War, and the quality of stirrers and pots was gradually improved by the use of better clays and by paying more attention to the composition of the refractory in relation to the glass composition and its corrosive action upon the pot.

The structure of the pot itself was greatly improved by adopting the process of slip casting. This method was developed by A. V. Bleininger and his colleagues at the National Bureau of Standards in the USA. It was a new

Table 3. National Bureau of Standards specifications

Constituent	Sand	Flint powdered	Boric acid, Borax, Saltpeter, Sodium Nitrate and Potash (83-85%)	Soda ash	Litharge	Barium carbonate and hydrate	Zinc oxide	Potash (99%)
	%	%	%	%	%	%	%	%
Fe_2O_3	<0.02	<0.025	<0.002	<0.003	<0.005	<0.005	<0.005	<0.0025
Cl	–	–	<0.10	<0.25	<0.10	<0.10	–	<0.125
SO_3	–	–	<0.10	<0.10	<0.10	<0.25	<0.35	<0.125
SiO_2	>99.25	>99.0	–	–	–	–	–	–
Al_2O_3	<0.20	<0.25	–	–	–	–	–	–
CuO_2	–	–	<0.001	<0.001	<0.001	<0.001	<0.001	<0.001
Cr_2O_3	–	–	<0.001	<0.001	<0.001	<0.001	<0.001	<0.001
MnO	–	–	<0.001	<0.001	<0.001	<0.001	<0.001	<0.001
CoO	–	–	<0.001	<0.001	<0.001	<0.001	<0 001	<0.001
NiO	–	–	<0.001	<0.001	<0.001	<0.001	<0.001	<0.001
MgO	–	–	–	–	–	–	–	–
$CaCO_3$	–	–	–	–	–	–	–	–
K_2CO_3								
Na_2CO_3	–	–	–	–	–	–	–	–
TiO_2	–	–	–	–	–	–	–	–
BeO	–	–	–	–	–	–	–	–
ZrO_2	–	–	–	–	–	–	–	–
BaO								
CaO and MgO	–	–	–	–	–	–	–	–

application of a process long practised in the pottery industry and it has now largely replaced the traditional method of pot-making, which is to build the pot up laboriously from the base, layer by layer, as described in Chaper 5.

The pot process for the melting of optical glass is intermittent and can only be used to produce glass on a relatively small scale. Although a method was developed for casting molten optical glass into a rectangular iron container and then cutting it up to form many small pieces suitable for lens blanks, it was not until the Second World War, under increased pressure of demand for glass, that the problem was tackled from the other end and a continuous melting process developed. After the war it was greatly improved and was introduced commercially in 1948.

The continuous optical process is very similar in principle to the manufacture of ordinary glass in a tank furnace, with the addition of very efficient stirring. It is shown schematically in Figure 22. The glass is melted, refined and homogenized in an electrically-heated platinum-lined tank, the refining process taking place at about $1400°C$ for flint and $1550°C$ for crown for batch materials used in the production of optical glass

Calcium carbonate	Barium nitrate	Lithium carbonate	Strontium carbonate	Titanium oxide	Beryllium oxide	Zirconium oxide	Arsenic and Antimony oxide	Alumina hydrate
%	%	%	%	%	%	%	%	%
<0.010	<0.005	<0.01	<0.030	<0.005	<0.020	<0.030	<0.025	<0.010
<0.10	<0.10	<0.10	–	–	–	–	–	–
–	<0.10	<0.10	–	–	–	–	–	–
–	–	–	–	–	–	–	–	–
–	–	–	–	–	–	–	–	>64.3
<0.001	<0.001	<0.001	<0.001	<0.001	<0.002	<0.003	–	–
<0.001	<0.001	<0.001	<0.001	<0.001	<0.002	<0.003	–	–
<0.001	<0.001	<0.001	<0.001	<0.002	<0.002	<0.003	<0.025	<0.005
<0.001	<0.001	<0.001	<0.001	<0.001	<0.002	<0.003	–	–
<0.001	<0.001	<0.001	<0.001	<0.001	<0.002	<0.003	–	–
<0.50	–	–	–	–	–	–	–	–
–	–	<0.50	–	–	–	–	–	–
–	–	<2.5	–	–	–	–	–	–
–	–	–	–	>98.5	–	–	–	–
–	–	–	–	–	>99.0	–	–	–
–	–	–	–	–	–	>98.5	–	–
–	–	–	<1.0	–	–	–	–	–

glass. After refining, the glass is homogenized by stirring with a platinum stirrer. The glass emerges from the tank to be formed into gobs by a feeder or as a thick bar or ribbon to be rolled, extruded as strips or panes, cast into blocks or moulded to form lens or prism blanks. These blanks, formed in an automatic pressing machine, may be compared with the continuously produced pressed glass articles of soda-lime-silica glass. Annealing is also continuous, the initial cooling rates increasing from less than $1°C$ per hour to $360°C$ per hour as the temperature falls. This annealing not only relieves strain but also controls the optical properties of the glass. The refractive index varies with the cooling rate and the annealing schedule must be designed so that every part of the glass cools through the same time-temperature cycle.

The continuous process produces glass far more rapidly, cheaply and with more control than the intermittent pot method, where melting, refining and homogenizing took about forty-eight hours and after slow cooling the pot full of glass was broken open to give, at the most, about twenty-five per cent of

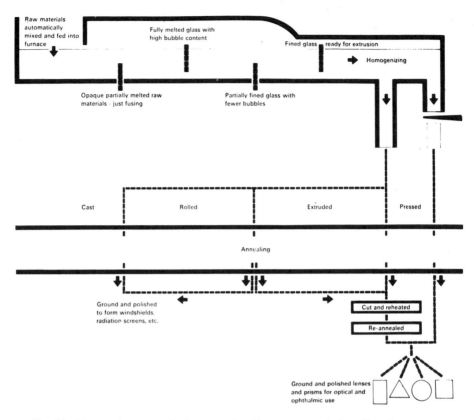

Fig. 22. The continuous optical process. In this process optical quality glass emerges continuously from the melting tank to be formed into panes, strips, blocks or lens and prism blanks.

suitable material. The use of a platinum-lined pot also removes contamination by solution of refractory.

The quality of the glass has also been improved by the increasing availability of pure raw materials and chemicals. Table 3 gives the American standards of purity for the major raw materials used in optical glass manufacture, as specified in 1948 by the National Bureau of Standards. The major contaminants are the iron oxides Fe_2O_3 and FeO and the N.B.S. figures specify that the sand used for optical glass manufacture must contain less than 0.02 per cent of Fe_2O_3. A British Standard of 1958 gives a maximum value of 0.008 per cent, which may be compared with the maximum permissible levels for container glass of 0.03 per cent. It is possible to add manganese or selenium as a decolorizer to counteract the colours of iron, but this is not desirable for optical glass because, whilst the addition of manganese produces a colourless glass, too much manganese can cause a reduction in overall transmission. Also, glass containing manganese may suffer from solarization, a process by which it acquires a purple colour when exposed to sunlight over a long period.

The chemical industry was also able to supply increasingly pure products. In the United States, which is now one of the world's leading makers of optical glass, the quality of barium and potassium salts, even as late as 1937, was poor and there was no domestic source of barium carbonate or potassium nitrate. However, one manufacturer had just started to make suitable barium hydroxide which could be substituted for barium carbonate, and satisfactory potassium carbonate was also being produced. By September 1939, when an emergency situation was declared, barium nitrate was the only essential ingredient not domestically available, but this situation was remedied in 1940, and shortly afterwards the manufacturers were able to supply very pure grades of zirconia, beryllium oxide, titanium oxide and the rare earth oxides, so laying the foundations for the spectacular development of the American optical glass industry.

Bibliography

1. *The making of optical glass in India: its lessons for industrial development,* Shanti Swarup Bhatnager Memorial Lecture, 1961, Atma Ram, *Proc. National Institute of Sciences of India,* 1961, **27A**, 6, 531.
2. The Bakerian Lecture: *On the manufacture of glass for optical purposes,* December, 1829, Michael Faraday, *Phil. Trans. R. Soc.,* 1830, **120**, (1), 1.
3. *Michael Faraday and Optics. Retrospect on the occasion of the 100th anniversary of his death,* Wilhelm Shütz, *Jena Review,* 1967, 315.
4. *Astronomical telescopes,* G. M. Sisson, *J. Soc. Glass Technol.,* 1951, **35**, 260.
5. *Die optischen Hilfsmittel der Mikroskopic,* E. Abbe, *Bericht Uber d. wissensch. Appar. auf d. Londoner intern. Ausstellung i. J.* 1876, **I**, 417 (Brunswick, 1878).
6. *History of optical glass production in the United States,* F. W. Glaze, *Bull. Am. Ceram. Soc.,* 1953, **32**, 242.
7. *A History of Chance Brothers and Co.,* J. F. Chance, privately printed by Spottiswoode, Ballantyne and Co. Ltd., 1919.
8. *A textbook of Glass Technology,* F. W. Hodkin and A. Cousen, Constable and Co. Ltd., 1925.
9. *Glass in the Modern World,* F. J. Terence Maloney, Aldus Science and Technology Series, 1967.
10. *Glass throughout the Ages,* R. J. Forbes, *Philips tech. Rev.,* 1960-61, **22**, 9-10, 282.
11. *The centenary of optical glass manufacture in England,* The Second Chance Memorial Lecture, H. Chance, *J. Soc. chem. Ind.,* 795, London, 1947.
12. *Jena glass and its scientific and industrial applications,* H. Hovestadt (trans. J. D. Everett and A. Everett), Macmillan and Co. Ltd., 1902.
13. *Carl Zeiss: on the 150th anniversary of his birthday,* H. Gause, Jena, 1966.
14. *Ernst Abbe: University teacher and industrial physicist,* W. Schütz, Jena, 1966.
14. *Telescopes from the Optical Museum of the Carl Zeiss Foundation,* Jena, *Jena Review,* 1970.

5

Refractories and furnaces

When sand and other materials are melted together at high temperatures to make glasses a very corrosive liquid is formed. The pots and furnace parts which come into contact with this liquid have to be made of high melting point refractory oxides which can resist this corrosive action. Since about 1930 increasing numbers of refractories have been manufactured with excellent properties but before then and from ancient times glassmakers relied on natural rocks and clays.

Natural clays consist mainly of silica and alumina together with some water and impurities such as iron oxide, alkali and alkaline earth oxides. When fired to a sufficiently high temperature the water is removed and the clay changes to a mixture of silica and mullite crystals $(2SiO_2.3Al_2O_3)$ together with a little glassy material depending upon the temperature of firing and the impurities present. Although successful selection of suitable clays and rocks had long been practised it was not until the early 1920s that Bowen and Greig, working in the Geophysical Laboratory in Washington, first produced data which provided a scientific foundation for the description of refractory materials. The information is summarized in Figure 23. In this figure a point along the base represents one particular ratio of silica to alumina; the temperature at which the various compositions are completely liquid is shown by the full line; between the full line and the dashed line the material consists of the crystal named with the remainder in the liquid state. The dramatic lowering of the melting point of silica by the first additions of alumina can be seen, but when the alumina content increases above about five per cent the crystal present becomes mullite and the temperature at which all the material is liquid increases until the melting point of mullite is reached at $1850°C$. The reason why good refractories are made from clays with a high alumina/silica ratio immediately becomes apparent. Unfortunately, the story is not complete as impurities lower the temperature considerably; as an example, a dotted line in the diagram indicates the effect of five per cent of sodium oxide. The resistance to attack by the molten glass is also much reduced if the fired material is porous and the porosity depends upon the nature of the raw clay and the heating, or firing, which has been given to the refractory before use.

Analyses of Egyptian crucibles of about 1370 BC suggest that the clay contained considerable amounts of alkali, lime, magnesia and as much as six to eight per cent of iron oxide; a pot made from this clay could hardly have been used above $1100°C$. While simple wood-fired furnaces were used it was

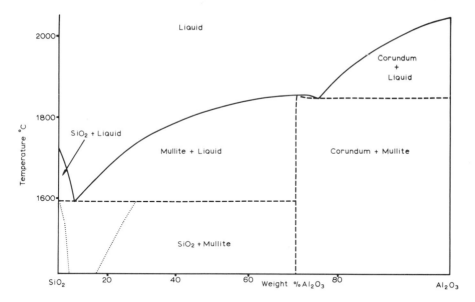

Fig. 23. The natural clays used by the old glassmakers to build their pots and furnaces were complex compounds of silica and alumina together with some water and impurities such as iron oxide and alkali and alkaline earth oxides. This diagram, known as a phase diagram, indicates the temperature at which a particular composition is completely liquid. For example, for a clay consisting of forty per cent alumina and sixty per cent silica, all the material is liquid above about 1780° C. The dotted line indicates the effect of an addition of five per cent of the impurity sodium oxide to the various silica-alumina mixtures; the melting point is lowered considerably and such a clay is much less refractory than one which is free from sodium oxide.

not possible to obtain much higher temperatures and not much change in the pots took place; analyses of eleventh-century pots from Kiev gave silica/alumina ratios of about three to one but several impurities were also present which would again suggest about the same melting temperature.

There are very few references to refractories in the seventeenth-century texts of Neri and Merrett, which are chiefly concerned with the preparation of glasses, but Merrett mentions that clays from Nonsuch in Sussex and from Worcestershire were generally used for pot-making; the latter was probably Stourbridge clay which later came into widespread use in Britain.

Some improvements had been made by the middle of the nineteenth century. Georges Bontemps, in his *Guide du Verrier* of 1868, made an important distinction between the materials to be used for the pots and for the construction of the furnace. The furnace was to be built of materials which could resist high temperatures and display minimum shrinkage on firing; materials which were to come in contact with molten glass had, in addition, to resist the attack of the glass. He also stated that refractories must be of high thermal conductivity; otherwise the glass would be of inferior quality, melting times would be increased and more flux would be required. In modern furnaces high thermal conductivity is not necessarily a desirable feature of refractories. Bontemps goes on to recommend a clay with minimum impurity content and a high alumina/silica ratio; alumina, he says, gives strength to the dried material:

> A glassmaker ought to limit himself (in the first place) to a simple chemical analysis, and then, after noting also the apparent qualities of the clay, he will be able to say whether it is necessary to make a large pot finally to test his conclusions.

The clay from the quarry was mixed with a quantity of crushed burnt fireclay, or grog, either in the form of specially selected fireclay burnt for this purpose or, to a lesser degree, cleaned fragments of old pots. The burnt material diminished the contraction which took place when the pot was dried and later fired, a contraction which was caused by loss of water held by the raw clay. The amount of burnt material added depended upon the nature of the clay, a very plastic, or fat, clay requiring an addition of up to fifty per cent, whilst for leaner or less pliable clays, such as those of the Stourbridge area, much smaller proportions were used. The greater the proportion of grog added the easier the drying process and the less fragile the pots during firing but the finished pots were more porous and more easily attacked by the molten glass. As the grain size of the grog or burnt material decreased, the pots became tougher, with reduced porosity.

Bricks for the walls and crown of the furnace, where shrinkage was critical, were made from a mixture of approximately one part clay to three parts grog but the proportion of grog to clay could be increased. If very pure sand was available this could be used instead of grog to produce refractory bricks which had the property of dilating slightly upon heating. These bricks were used for the crown of the furnace, as the joints did not open to give surfaces where the flames could attack and produce droplets, which then fell into open pots or tanks of molten glass. As a small amount of dropping was unavoidable, the high purity of the crown bricks was necessary to prevent serious discoloration of the melt. Natural stone could be used if the composition was suitable but Bontemps states that manufactured bricks were found to be infinitely superior.

Bontemps' instructions are completely in accord with the data produced in laboratories some seventy years later. The insistence on very pure sand for silica bricks should be noted as a requirement entirely consistent with the sharp fall in the melting point of silica on adding alumina or alkali. However, some impurities do not have such initial effects: if, for example, only lime were present the melting point of silica is very little changed until thirty per cent of lime has been added. Nevertheless, from the frequent references we have made to Bontemps it is clear that he was an outstanding glass technologist.

Pot-making

The Egyptian crucibles were very small, measuring only a few inches in depth and diameter but the discovery and growth of the art of glass-blowing made larger pots necessary. The craft of pot-making has not changed much through many centuries. The pot-maker today uses the traditional methods but he is aided by being able to work in pot-rooms with controlled temperature and humidity so that the drying conditions can be optimized.

The plastic mixture of raw clay is allowed to stand for several months in order to attain the best working consistency and is thoroughly compressed and mixed to remove air bubbles, either mechanically or still sometimes by the traditional method of treading the clay with the feet. The pot-maker handles the clay in rolls about four inches long and two inches in diameter, which he throws on to a circular board. He works the clay with his hands until the required thickness for the bottom of the pot is reached. The board is then lifted, turned over and removed from the clay bottom by means of a wire or band saw, the bottom then being allowed to stand for a while before building the side walls. A rim is pushed up around the edge of the bottom and on this rim the side wall is built from rolls of clay. The pot-maker moves around his pot supporting the inner surface and adding the clay rolls. He strokes his fingers along the clay to spread it, break up any possible air bubbles and facilitate drying of the surface before the next layer is added. The pots are built in batches and each is built upwards in sections, one section being allowed to dry somewhat before the next section is added, in order to support the weight of the new section.

If the pot is a covered or 'caped' pot, the pot-maker must now gradually build the walls inwards and close the top of the pot before inserting a wooden pattern in the side to form the pot mouth. The rim of the mouth is then built round the pattern shape, and the inside of the pot mouth is cut away when the pattern has been removed. Figure 24 shows pot-makers at work.

The completed pots are left to dry for some months but before use they have to be brought up to temperature slowly. This pre-heating is done in a separate furnace, or pot-arch, and the pot transferred to the working furnace at bright red heat, an operation known as pot-setting.

In more recent years pots have been made by slip casting, a technique, long established in the pottery industry, of preparing a suspension, or slip, of the clay and pouring this slip into a plaster mould which absorbs water until a layer of more or less solid clay forms on the wall of the mould. When this layer is thick enough the remaining slip is poured out and after drying the pot is ready for firing. A modified process of slip casting includes pressing. A very fluid slip is poured into an outer plaster mould, a plaster former is inserted into the mould and applied pressure pushes the slip upwards between the inner and outer moulds. After the slip has set the plaster former is removed and the pot is ready for firing. Closed pots are made in two parts and then

Fig. 24. The hand manufacture of glass melting pots. In the background the pot maker has built up the walls of the glass pot layer by layer and is now turning in the walls preparatory to closing the top of the pot. In the foreground the mouth of the pot has been built around a wooden former and is receiving a final smoothing after the removal of the former.

joined together by an application of slip to the join. This method of pot manufacture is widespread in the USA but most British pots are still made by the traditional methods.

Pot-melting remains now almost entirely in the high-quality handmade crystal glass industry but nearly all glass melted today is made in tank furnaces. In these tank furnaces the glass is contained in a huge bath which may have a capacity of up to 2000 tons of molten glass. The furnaces incorporate new refractories developed during the last fifty years. But before discussing the development of these materials it is appropriate to deal first with the history of glass furnaces.

The first furnaces

From the earliest times glassmakers carried out their operations in two or three separate furnaces, or in separate compartments of one furnace structure. Often the heating structure was divided into compartments, clearly in an attempt to use, for secondary purposes, heat that would otherwise have been wasted from the melting furnace. The successful recovery of waste heat so that it could be passed back to the melting chamber was not accomplished until the invention of the Siemens furnace in 1856.

The earliest descriptions of glass furnaces are contained in a series of Syriac and Arabo-Syriac manuscripts, now deposited in the British Museum. One, which dates from the ninth century BC or later, contains the following description: 'The furnace of the glass-makers should have six compartments, of which three are disposed in storeys one above the other. . . . The lower compartment should be deep, in it is the fire; that of the middle storey has an opening in front of the central chambers, these last should be equal, disposed on the sides and not in the centre, so that the fire from below may rise towards the central region where the glass is and heat the materials. The upper compartment, which is vaulted, is arranged so as uniformly to roof over the middle storey; it is used to cool the vessels after their manufacture.' A smaller furnace is also mentioned for fritting, a process in which the raw materials were raked over from time to time to expose fresh surfaces to the flame. This process eliminated some of the gaseous products from the melting together of the batch materials and burnt off carbonaceous impurities in the fluxes.

Much of the very early glass, especially that made in Egypt, is opalescent or opaque and contains many bubbles, indicating that the furnace temperatures did not reach much more than 1100°C, as was estimated from the composition of the refractories. Temperatures probably rarely exceeded 1200°C until the eighteenth century.

The earliest picture of a glass furnace was found in a manuscript of AD 1023 from the monastery of Monte Cassino. The manuscript is thought to be a copy of an original by Rabanus Maurus, Bishop of Mainz, who died in AD 856. The picture (Figure 25) is therefore to be ascribed to the ninth century AD. Although it shows the three compartments, one for the fire, the central compartment for the melting pots and the upper chamber for the slow cooling of finished articles, it can hardly be said to give sufficient detail to show how the exquisite beauty of the Portland Vase and the Lycurgus cup could be created.

By the tenth century the furnaces used in northern Europe, although showing considerable variation, had a remarkably different pattern from those used in the Mediterranean countries. A typical Northern furnace was described in the twelfth century by the German monk Theophilus in his book *Schedula Diversarum Artium.* A model of the furnace in the Science Museum, London, is shown in Figure 26. The overall dimensions given by Theophilus are fifteen feet by ten feet by four feet.

The fire burnt under the platform, or siege, which formed the base of two compartments; the small one on the right was used for fritting and the larger one had four openings on each long side to give access to the pots. Holes in the siege allowed flames and heat to come up from the firing chamber. Nothing is said about pre-heating the pots but a separate furnace eighteen feet by eight feet by four feet was used for cooling the ware, a process now known as annealing.

HG—8

Fig. 25. A medieval glass furnace—an illustration taken from an eleventh-century manuscript. The man sitting on the three-legged stool is blowing a vessel in front of the glass furnace. The furnace is built in three parts; in the lower part is the fire, in the middle the melting pots, and in the upper part the vessels are slowly cooled. Three openings for working the glass can be seen in the middle compartment; they also act as exits for the escape of smoke and fumes.

Fig. 26. A model of a glass furnace according to the description of Theophilus. This rectangular furnace is typical of the type of furnace used in northern Europe during the medieval period. The heat from the fire in the lower compartment rose through the floor, or 'siege', and heated the glass melting pots in the upper compartment. The right hand section of the upper compartment was used for pre-heating the batch before melting it in the pots, a process known as 'fritting'.

At Vann Farm in Chiddingfold, Sussex, the remains of a rectangular Tudor furnace have been found which appears to have had wings for fritting, pre-heating the pots and annealing.

A Bohemian forest glass house of the fifteenth century is shown in Figure 27, an illustration from a Bohemian manuscript now in the British Museum. The furnace is rectangular, the typical Northern shape, and it

Fig. 27. A Bohemian forest glass house of the fifteenth century. In the background of the picture a man is digging sand from the hillside, and flux and fuel are being carried in sacks and baskets. In front of the furnace are the glass blowers gathering glass and blowing a vessel, whilst a boy tends the furnace, and the worker on the left removes the vessels for annealing.

probably has an annealing furnace added on one end. The picture tells the full story of forest glass production; a man is digging sand from a hill with a stream at its foot, another carries fuel in baskets and flux in the form of ashes in a sack. The boy is tending the furnace fire while the three craftsmen gather glass, shape a vessel and remove the finished articles for the annealing treatment. The German scholar, Georgius Agricola, in his great book on metallurgy and related subjects *De re metallica* published in 1556, described what has since become known as the 'German' furnace (Figure 28). Agricola described several systems of melting involving one, two or three furnaces, but the processes of fritting, melting and annealing were common to all systems. In the furnace illustrated the upper section of the melting furnace was used for annealing, although no entry to it can be seen in the picture. These beehive furnaces were first used in Mediterranean countries and are sometimes referred to as southern furnaces.

None of the descriptions of the old furnaces has anything explicit to say about the provision of draught to draw air through the furnace. The old writers Theophilus and Agricola were reporting observations and appear to have missed the importance of a draught; even Agricola's drawings omit an opening at the top of the furnace. Although the old glassmakers did not understand the process of combustion, their empirical knowledge often led them to site their furnaces in such a way that the prevailing winds provided a good draught.

The introduction of coal-fired furnaces

Making a virtue out of the necessity for using coal instead of wood, the English glassmakers soon developed furnaces that could exploit the superior properties of coal as a fuel. In the early seventeenth century Paul Tyzack, possibly the first glassmaker ever to use coal with complete success, brought to Stourbridge considerable skill in the art of glassmaking which, combined with the great interest in this area in coal-firing for the glass and iron industries and the availability of excellent clay for furnace manufacture, ensured the success of his enterprise.

Until the seventeenth century glass houses were built like barns with pitched roofs and usually a small turret above the furnace to let out smoke and fumes. During the late seventeenth century the 'English' or cone-shaped glass house was introduced, permitting the concentration of all air currents in a single upward movement with improved ventilation, much higher furnace temperatures and more efficient use of fuel. These furnaces burnt coal, but the use of coal presented several difficulties. Greater care had to be taken in the choice of clay for pot materials and it was found necessary to 'close the pots', that is, to use covered pots to protect the glass from smuts and sulphurous fumes given off by the coal. The introduction of covered pots was a very important factor in the development of English lead glass which must be protected from direct contact with the flame.

A—Blow-pipe. B—Little window. C—Marble. D—Forceps. E—Moulds by
means of which the shapes are produced.

Fig. 28. A southern glass furnace of the sixteenth century. The furnace has three
sections, the lower one for the fire, the middle one for the glass pots, and the upper one
for annealing the glass. The glass blowers are working around the furnace and the vessels
are packed in the large box seen in the bottom right hand corner. In the background the
sale of the glass is being discussed, and a pedlar carries away the vessels. In those days the
customer either ordered his glass direct from the maker or bought it from the travelling
hawker.

The English cone furnace

Coal furnaces spread throughout the country during the late seventeenth and early eighteenth centuries and towards the end of this period the tall cones of the 'English' glass house became a familiar sight. The largest ones rose to over ninety feet and were thirty to forty feet or more in diameter at the bottom. Figure 29 shows a glassworks at Lemington-on-Tyne in about 1912. The large

Fig. 29. Glass cones at Lemington-on-Tyne, 1912. These cones were often over 100 ft high; the cone in the background was built in 1789 and is now preserved as a building of great historical interest.

cone at the back of the photograph is one hundred and twenty feet high; it was built in 1789 and is now preserved as a building of great historical interest. The work going on inside a cone is shown in Figure 30, this etching makes a pair with Figure 4 and gives a clear view of the operations of gathering, blowing, marvering and working at the chair. A pot-arch is indicated at the back of the picture. The building of such cones to house the new round furnaces was not without its problems. Throughout the eighteenth century, newspapers frequently reported the collapse of these structures, most of them occurring without warning.

The furnace inside the cone was direct-fired from a fireplace in the middle below ground level and air was supplied to the fire via an underground tunnel. The flames rose into the furnace and passed over the pots; combustion products escaped through flues in the side walls and so up through the cone to the outside air. The cone itself performed the function of a tall chimney in

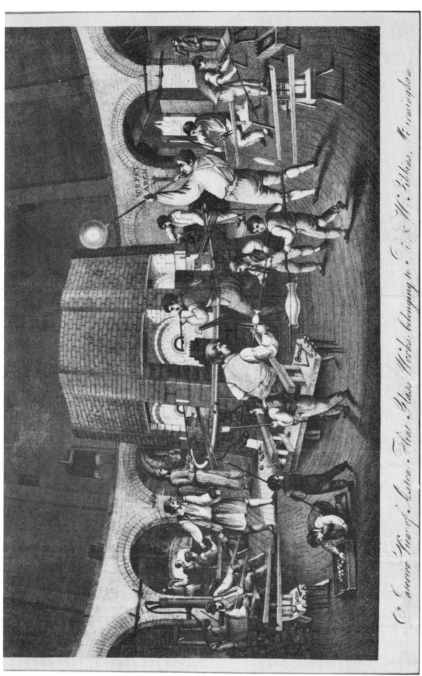

Fig. 30. The interiors of the glass cones were spacious and provided good working conditions for the glass blowers. Around the walls of the cone can be seen the mouths of the annealing ovens and the 'pot arches', or ovens for pre-heating the pots before subjecting them to the fierce heat of the furnace.

increasing the draught. The roof, or crown, of the furnace was suitable for drying pots, prior to pre-heating in the pot-arches at the sides of the furnace, and fritting was also carried out in a small furnace adjacent to the melting furnace inside the cone. An additional and important advantage of the cone furnace was the greatly improved comfort of the working conditions. Figure 31 shows a plan view of an English cone furnace from a French encyclopedia written by Diderot and d'Alembert and published in the mid-eighteenth century.

Developments in wood-fired furnaces

Not all furnaces used coal as fuel. Until the middle of the eighteenth century both plate and sheet glass were founded in furnaces fired with wood and, in the absence of legislation prohibiting the use of timber as a fuel, wood-firing continued in France for much longer than in England. Following the vast investment in the flat glass industry under the patronage of Louis XIV and his minister Colbert, an advanced and expensive plate glass plant, the *Manufacture Royale des Glaces,* was established at St Gobain in 1693, but it was not until 1763 that the glassmakers of St Gobain attempted, without success, to substitute coal for wood and in 1819 they were still using wood. By 1829, they had succeeded in melting in a coal furnace but still had to fine, or remove gas bubbles, in a second, wood-fired, furnace. In England, the manufacture of cast plate glass by the French method was not a commercial success until the end of the seventeenth century, mainly because the French workers who tried to introduce the process were unaccustomed to the English coal-burning furnaces. In 1792, Robert Sherbourne was appointed to manage the Ravenhead Plate Glass Works in Lancashire. He soon introduced covered pots which protected the glass, a very important factor in the eventual successful manufacture of plate glass at Ravenhead.

Plate glass furnaces were built on a large scale, as can be seen in the cut-away drawing of Figure 32 taken from the encyclopedia of Diderot and d'Alembert, which shows how the pots were placed in the furnace. The main furnace was situated in the centre of the hall, with annealing ovens ranged down each side, their floors level with the casting tables. Figure 33 is a plan view of the furnace. On a lower level, air entered the furnace via a tunnel. The cuvettes could be withdrawn from the furnace through larger openings directly below the openings shown in the plan.

Improvements in direct-fired furnaces during the nineteenth century

In all the furnaces described so far, the fuel was burnt and the products of combustion allowed to pass directly over the pots—these are known as direct-fired furnaces. Improvements in direct-firing were introduced in the late nineteenth century; for example the Frisbie feed was a mechanical device

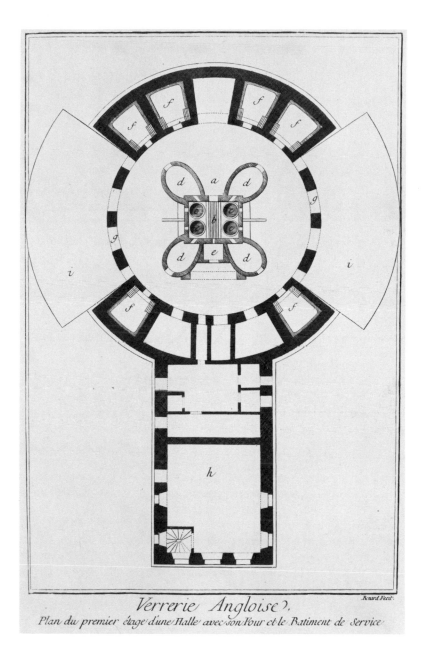

Verrerie Angloise.

Plan du premier étage d'une Halle avec son Four et le Batiment de service

Fig. 31. Plan view of an English cone furnace: (a) main oven; (b) grating for coal; (c) pots in the oven; (d) pot arches for firing pots before placing them in the oven; (e) fritting oven; (f) small annealing furnaces.

Fig. 32. Pots being placed in a plate glass furnace. The pots were very large and heavy and levering them into position was a difficult operation. Molten glass from these pots was therefore transferred to smaller containers which could be carried to the casting table.

which is said to have fed the fuel upwards into the fire instead of placing new coals on top of the existing fire. This upward feed caused the fuel to be partially gasified; the gases passed upwards and burnt in the furnace. Better combustion and higher temperatures resulted but the fuel consumption was only slightly reduced. Eventually the generation of gas was separated from the furnace and arrangements were made for some of the heat contained in the hot products of combustion to be taken up by the incoming gases. An early step towards achieving these great improvements can be seen in the Boetius furnace, a semi-direct-fired furnace which was widely used in Europe and Great Britain in the latter part of the nineteenth and well into the twentieth centuries. The furnace was invented around 1865-70 but the method of pre-heating, now known as recuperation, was not widely used until the twentieth century. A cross-section of the furnace is shown in Figure 34.

Coal was fed in at the sides of the furnace onto the fire grates and combined with air, known as primary air, to produce a combustible gas by a process of incomplete combustion. The air needed for burning this gas, termed secondary air, passed upward through channels around the fire-boxes or gas producers, thus undergoing a slight amount of pre-heating. The gas and secondary air then passed into the furnace and burnt together. The design was a slight improvement on the old direct-fired furnace, using on average 1.5 to 1.75 tons of coal per ton of glass, against the previous 2.0 to 2.5 tons. The

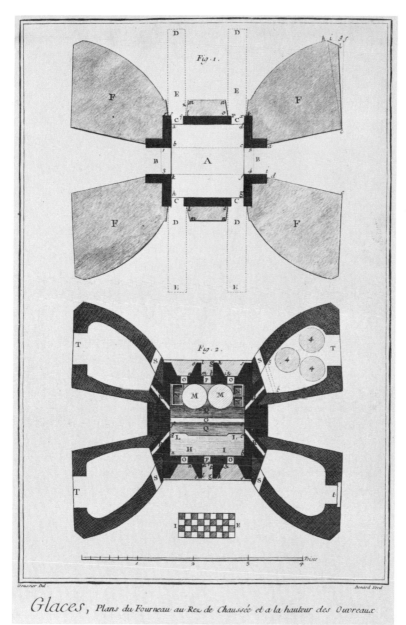

Fig. 33. Plan view of a plate glass furnace. H,I. siege for the pots; L. siege for the 'cuvettes', or rectangular clay containers into which the glass from the pots was ladled before removal from the furnace; N. cuvettes; M. pots in position; G. space between the sieges; O. openings through which glass was transferred from pots to cuvettes for pouring; T. pot arches for firing the pots; t. side furnace where batch was treated before remelting in the main furnace.

pre-heating of the secondary air, although small, helped considerably to burn the gases, thereby reducing the amount of unburnt gases in the stack and saving fuel. Unfortunately if the pot broke the contents ran down and blocked up the space, or eye, through which the furnace was fired.

The name most associated with the development of the modern glass melting furnace is Siemens. There were five brothers active in many European countries in designing furnaces; one of the brothers, C. W. Siemens, acknowledged very clearly his debt to the contemporary investigations which showed that heat and energy were interchangeable.

Advances in the understanding of the nature of heat

As the eighteenth century drew to a close, great advances were being made in the physical sciences. It was increasingly being realized that theories must be supported by experimental measurements, and this quantitative checking of theories produced amongst other things, a realization that heat and energy were equivalent. Earlier ideas suggested that heat was an invisible, weightless, elastic fluid, termed caloric, which material bodies could absorb, their temperatures thereby being raised. Different parts of this fluid were supposed to repel one another and so caloric would be forced out of a hot body into a cold body when they were placed in contact. To account for differences in the specific heats of different bodies, matter was supposed to possess an attraction for caloric which varied with the chemical constitution of the body.

These ideas were found to be in error by Count Rumford in a series of experiments performed in 1799. Whilst engaged in superintending the boring of cannon at the naval arsenal at Munich, the Count was impressed by the high temperature of the metallic chips thrown off. The calorists explained this observation by suggesting that during the boring the particles of the bodies were pressed closer together, thus extruding caloric; further, that the specific heat of the substance in fine division was less than that of the same substance in large masses. Rumford pressed a blunt borer against one end of a cylindrical mass of gun-metal which could be rotated by a horse-drawn device; a flannel covering prevented heat loss. After 960 revolutions had been made, the 113 lb of gun-metal had been raised in temperature from $60°F$ to $130°F$, with the production of less than half an ounce of metallic dust. Rumford questioned whether so much heat could have been furnished by so small a quantity of dust, and showed by direct experiment that the specific heat of the metal employed was sensibly the same whether it was in the form of dust or of a large mass. Because the evolution of heat, as measured by the rate of rise of temperature, was as brisk throughout the experiment as at the beginning, he concluded that there was no limit to the amount of heat that could be produced by friction, and therefore that the heat could not be present as a fluid within the body.

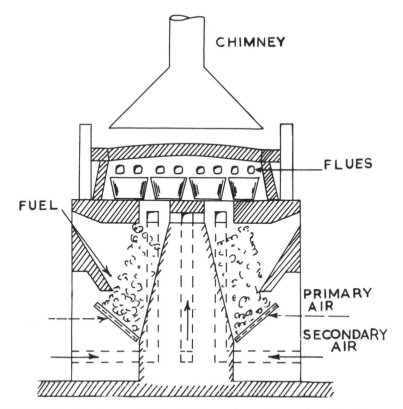

CHIMNEY

FLUES

FUEL

PRIMARY
AIR

SECONDARY
AIR

Fig. 34. The Boetius furnace. Coal and primary air, entering as shown by the arrows, burned incompletely to give a combustible gas. Secondary air to burn this gas passed upwards around the fire-boxes, thus undergoing slight pre-heating. The gas and secondary air then passed into the furnace and burned together. The Boetius furnace, by utilizing some of the heat of combustion, was more efficient than the direct-fired furnace and used less fuel to melt the same amount of glass.

Humphry Davy, in the same year, also showed that the concept of heat as a material or quasi-material fluid must be abandoned. He rubbed two pieces of ice together and found that after a short time, nearly all of the ice had melted: he concluded that any quantity of ice can be melted by a sufficient amount of rubbing, and thus confirmed Rumford's statement that an unlimited supply of heat can be obtained by rubbing two bodies together for a sufficient length of time.

The experiments of Rumford and Davy involved friction between bodies maintained in relative motion. The question now arose as to whether there was any relation between the work performed and the heat produced. The constant relationship was established by the experiments of Joule, commencing in 1840. His apparatus consisted of a paddle wheel which could be revolved in a tub of water by a system of cords and weights. The weights were wound up and then allowed to fall a fixed distance from which the work done on the water by the paddle wheel as it revolved could be calculated. The heat generated was found by measuring the temperature of the water before and after the experiment. Joule found that every time he made the experiment, a constant amount of heat was produced by a given amount of work. His measurements were the basis for the First Law of Thermodynamics, a statement of the observation that when work is transformed into heat, or heat into work, the quantity of work is equivalent to the quantity of heat. This concept of equivalence between work and heat led very quickly to improvements in furnaces.

The regenerative principle

C. W. Siemens wrote in a paper published in the Proceedings of the Institution of Mechanical Engineers in 1857, that:

> Our knowledge of the nature of heat has been greatly advanced of late years by the investigations of Mr J. P. Joule of Manchester, and others; which have enabled us to appreciate correctly the theoretical equivalent of mechanical effect or power for a given expenditure of heat. We are enabled by this new dynamic theory of heat to tell, for instance, that in working an engine of the most approved description we utilize at most only one-sixth to one-eighth part of the heat that is actually communicated to the boiler, allowing the remainder to be washed away by a flood of cold water in the condenser. If we investigate the operations of melting and heating metals, and indeed any operation where intense heat is required, we find that a still larger proportion of heat is lost, amounting in some cases to more than 90 per cent of the total heat produced.
>   Impressed by these views the writer has for many years devoted much attention to carrying out some conceptions of his own for obtaining the proper equivalent of effect from heat. . . . The regenerative principle appears to be of very great importance and capable of almost universal application.

C. W. Siemens had indeed put much effort into obtaining the 'proper equivalent of effect from heat'. Amongst his earlier inventions were a regenerative steam engine and condenser, a regenerative evaporator and an apparatus for the economic production of ice. He also worked with Carl Lorenz, a Viennese bronze founder, during the 1850s experimenting on a new founding process, and he said that the experience that he gained during this period on the gas heating of furnaces put him on the trail of his later discoveries. The Austrians were at this time pioneers in gas heating and developed various furnace heating systems.

In 1856, British patent 2861 was granted to his brother F. Siemens for the invention of an 'Improved Arrangement of Furnaces, which Improvements are applicable in all Cases where Great Heat is required'. The invention was described in the following words:

> My improvement consists in so arranging smelting and heating furnaces, smith fires, etc., that the products of combustion on their passage from the place of combustion to the stack or chimney shall pass over an extended surface of brick, metal, or other suitable material, imparting heat thereto, which heat serves to heat the atmospheric air or other materials of combustion. The result of this arrangement is, that (they) are nearly heated to the degree of temperature of the fire itself, in consequence whereof an almost unlimited accumulation of heat or intensity may be obtained.

Four chambers were constructed containing brickwork with channels through which gases could pass. Through one pair of these chambers the products of combustion were passed to their final exit through the chimney. These hot gases heated the brick work and after twenty to thirty minutes a reversal valve caused the combustion products to pass through the second pair of chambers while the incoming gas and air each passed through one of the first pair of chambers. The cycle continued, the heat of the combustion products was said to be regenerated and the chambers were known as regenerators.

The principle of regeneration was applied to the re-heating of iron and steel at the firm of Marriott and Atkinson in Sheffield in 1857. In the paper by C. W. Siemens already quoted, he stated that one such furnace had been in constant use for nearly three months, working quite satisfactorily and using only twenty-one per cent of the fuel required by the old furnace in heating the same quantity of metal. The furnace had also been applied to the puddling of iron in Bolton, but there was no mention at this stage of its use in the glass industry.

In Frederick Siemens' original design for a regenerative furnace the fireplaces for gasification of the fuel were built into the furnace: smoke and dust from the burning fuel were taken through the regenerators and into the furnace with deleterious effects on the glass. These defects were overcome by the introduction of gas-firing.

The introduction of gas producers

In 1861, British patent 167 was granted to C. W. and F. Siemens for the invention of 'Improvements in Furnaces'. This patent contains a discussion of the application of the principle of regeneration to glass melting and incorporates the important development of units for combustible gas production separate from the main body of the furnace:

> It is an essential part of our Invention that the solid fuel used, such as coal, lignite, peat etc., should be decomposed in a

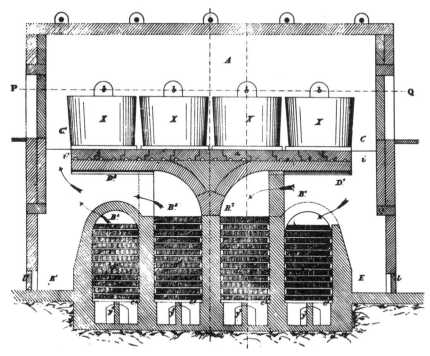

Fig. 35. The Siemens regenerative furnace, as applied to the manufacture of glass: longitudinal section. B^1, B^2, B^3, B^4 regenerators filled with checker-work of fireclay bricks; a. furnace floor; X. pots; b. ports in furnace for working pots; C^1, C^2, C^3, C^4 gratings connecting regenerators to the apertures, F, at their bases. These apertures are shown in the figure opposite.

separate apparatus so that the introduction of solid fuel into the glass furnace may be altogether avoided, and the gaseous fuel may be heated to a high degree prior to its entering into combustion with atmospheric air, also heated to a high degree, thus causing great economy of fuel. There is also great advantage derived from the absence of any solid carbon or ashes in the working chamber of the furnace, by which we are enabled to carry on operations in the open furnace which it has only been possible hitherto to conduct in covered vessels or pots. We are thus enabled to melt flint, extra white and other superior qualities of glass in open pots . . . or to melt steel and other substances upon an open hearth or bed without injury.

Figure 35 shows a longitudinal section of their improved regenerative furnace and Figure 36 shows a second longitudinal section through the connecting passage system between the furnace and the gas generators. The butterfly reversing valve is shown at g; in the position shown the gas is entering through the regenerator at the extreme left and the air through the next regenerator while the burnt gases are heating the right hand pair of regenerators.

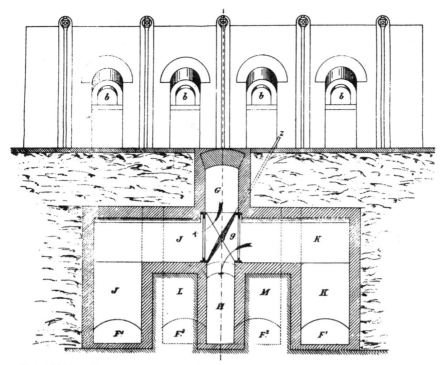

Fig. 36. The Siemens regenerative furnace, as applied to the manufacture of glass: longitudinal section through valve and passage system. F¹, F⁴, apertures connecting regenerators B¹, B⁴ alternately with the passage, G, from the gas producers, and the chimney passage, H, via a reversing damper, g. F², F³, apertures connecting regenerators B², B³ alternately with an air passage, I., and the chimney passage, H, via a reversing damper.

The gas was made in a separate unit known as a producer and the gas was called producer gas. The fuel was fed into the producer down a slope onto an inclined fire grate where it accumulated to a considerable thickness. The fireclay roof of the chamber radiated heat on to the fresh fuel causing partial combustion and volatilization. Air coming into the chamber through the bottom of the fuel bed resulted in a layer of incandescent coke and near the grate the coke burned to give carbon dioxide. Further up in the fuel bed the oxygen available was limited and the carbon burned to give carbon monoxide, the carbon dioxide reacted with more carbon and was reduced to carbon monoxide.

Experiments on the utilization of gaseous fuels had been made since the end of the eighteenth century. Bischof built a producer in 1839 which was identical in function to the Siemens producer but, perhaps because of the lack of the regenerative or recuperative principle at that time, it did not become a commercial success. Producer gas is still in use in the glass industry but it has almost entirely been replaced by natural gas in the USA, by oil and increasingly by the recently discovered natural gas in Europe.

The Siemens furnace in the glass industry

The regenerative system was tested during 1860 and 1861 in the glassworks of Lloyd and Summerfield in Birmingham and during the same period a regenerative furnace was under trial in the works of Chance Brothers in Smethwick. Charles William Siemens had been acquainted with the Chance family since the 1850s when Siemens had come to Birmingham to sell a silvering process. Robert Lucas Chance was very impressed by Siemens' inventiveness and when, ten years later, Siemens was ready to apply his ideas to melting glass he carried out much of his development work in Smethwick. An account of these trials was given in the discussion of a paper presented by C. W. Siemens at the Midland Institute, Birmingham in January 1862. J. T. Chance said:

> ... the regenerative furnace had been tried at Messrs Chance's glass works, and it certainly bid fair to produce a considerable change in the mode of maintaining the high temperature required in glass works. He was not in a position to state definitely that success was thoroughly obtained in every particular, simply because it required time to develop all the difficulties or peculiarities that might arise in the application of the furnace to a new process of manufacture, which might not be anticipated prior to actual trial. No difficulties however had occurred yet in the working of their large melting furnace, containing eight large pots holding two tons of glass each, which had now been in regular work for three weeks with complete success. ...

He reported on a small furnace which had been on trial for a year:

> ... In this first furnace some deposit had occurred in the regenerators, which he hoped would be obviated in the larger furnace. ... The volatilized gases (from the melted batch) would all have to pass through the regenerators and he was desirous of seeing whether this would cause any trouble by choking the passages of the regenerators after working for a length of time. ...

Dr Lloyd, of Lloyd and Summerfield, described their experiences as follows:

> ... they had one of the regenerative ten pot furnaces in operation nearly twelve months for flint glass making ... he was so much struck with the soundness of the principle that he went at once to see a small glass furnace that was working on that plan in Yorkshire; and being satisfied of the theoretical perfection of the plan, he adopted the new furnace immediately at his own works for flint glass making. ... He had adopted the new furnace mainly with a view to saving in fuel. ... It was built of about the same capacity as an old ten pot furnace, which was heated with large best coal ... the consumption was very considerable. The

result of comparison between the two furnaces was that the old furnace consumed as nearly as possible double the quantity of fuel required in the regenerative ... the coal used in the new furnace cost only one third as much per ton, being entirely small coal ... so that the actual cost of fuel in the new furnace was reduced to one sixth of that in the old, doing the same amount of work.

The Siemens regenerative furnace made it possible to use much higher temperatures and provided very substantial fuel economy; it still remains the core of modern glass manufacture. A large modern regenerative furnace melting glass for containers requires about 100 therms for melting a ton of glass, whereas the traditional non-regenerative furnaces of 1838 required 1800 therms per ton.

Initially there were problems in the use of the regenerative furnace, as Chance Brothers found in their trials during the 1860s. After a few weeks of operation the roof of the furnace began to drip under the influence of intense heat and the drops became so numerous that they were obliged to renew the roof, or crown, of the furnace. This fault was eliminated when a new crown was made of 'Welsh bricks, only one course in thickness so that the bricks could expand and the crown rise as the furnace got hotter and subside as it cooled'. Soot accumulated in the flues leading from the gas producers and it was found necessary to clean out the flues weekly. The problem still exists in producer-gas-fired regenerative furnaces, where it continues to be necessary to burn out the flues at regular intervals. Air leakage through the walls of the regenerators had to be overcome and it was found difficult to control the quality of the producer gas.

Day tanks

In a pot furnace, the melting chamber is only inefficiently utilized. The fuel is burnt outside the pots and heat must be transferred to the melt through the pot walls. Experiments were apparently begun in the early 1840s to eliminate pots, although the concept of a tank furnace first appeared in a British patent of 1769 granted to R. and R. Russel (BP 929); in this furnace the glass was to be contained in two small tanks with the fire between them. A patent taken out by Joseph Crosfield of Warrington in 1840 can be regarded as a transition from pot to tank furnace; flames passed over the batch, which was charged onto a sloping furnace floor, and the molten glass ran down into a pot. This furnace was never operated on a commercial scale.

The term 'tank' had been used in glassworks long before the invention of the tank furnace. In 1835, in the *Technological Encyclopedia* of J. J. Prechtl, a tank was an elliptical or rectangular pot, and in 1854 the term was applied to pots used for the manufacture of plate glass.

The Siemens' idea of a tank furnace for glass melting appears to have developed from the building of a water-glass furnace for the chemical factory of Wagemann, Seybel and Co., at Liesing near Vienna, in 1858. Frederick Siemens says that this furnace was 'for melting water-glass on the hearth, being a so-called tank furnace in contrast to a pot furnace'. Frederick left for England in 1859, and in 1860 he and his brother Hans (in Germany) separately experimented with tank construction. In the winter of 1860-61 Frederick built his first tank furnace: this would nowadays be described as an intermittent or day tank. Batch was melted at night and worked during the day. A trough-shaped floor was used as a container for the molten glass but, as a refining process was still lacking, the unrefined glass was run off and remelted in ordinary pots. Letters written by the Siemens brothers at this time form the earliest authority for the use of the new term 'tank furnace'.

Whilst Frederick Siemens worked in England, Hans Siemens did a great deal to develop and improve the intermittent glass tank in Dresden. He died in 1867 and the glassworks in Dresden were taken over by Frederick. Stimulated by his success with a continuous pot (see below) he succeeded, on 1st October 1867, in converting the intermittent tank to continuous operation for the first time.

Continuous tanks

Melting in pots is a process in which the temperature changes with time whilst the material remains stationary. The mixture of raw materials, the batch, is shovelled in until the pots are full, the first filling melts down and leaves space for the addition of more batch, sometimes a third 'filling-on' is necessary. The temperature is then raised for melting and refining; when the melt is in a satisfactory condition the temperature is lowered, perhaps by 100°C, to allow the glass to reach a satisfactory condition for gathering. In continuous tank melting the flames are adjusted to give a maximum temperature (hot spot) about two-thirds of the length of the tank from the position where the batch is filled-on. The batch piles float down the tank and disappear before the hot spot is reached. The glass which is flowing through the tank now enters the refining-zone. It then flows through a small opening, the throat, at the floor, or siege, level into the working end where it is conditioned for feeding to the machines by reducing the temperature. The melt thus passes through the stages of melting, refining and conditioning as it flows through the tank.

Frederick Siemens was thinking about continuous melting in the 1860s when he produced a continuous melting pot based on the observation that the density of molten glass increased as the melting process progressed. The pot consisted of three chambers connected by channels as shown in Figure 37. The spout, c, projected out of the furnace wall, and the refined glass was gathered here for working. The two uncovered sections, A and B, served for melting and refining respectively. Raw batch was charged into A.

As melting progressed, the denser, most completely melted glass accumulated at the bottom of A and rose through the channel, a, into the upper part of B. The process was repeated in B as the glass refined. As the lower part of the pot was cooler than the upper part, B served two purposes, namely for refining in the upper part and for cooling down the glass sufficiently for working in the lower section. The glass flowing into C via channel b was thus ready for working out. These pots worked quite satisfactorily for all kinds of glass, but they were never developed on a large scale owing to the high costs of manufacture.

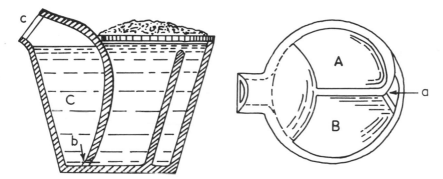

Fig. 37. The continuous melting pot–sectional and plan views. The batch is charged into chamber A and as it melts it sinks to the bottom of the chamber. As more batch is added, the molten glass is forced through an opening in the bottom of A and up channel a into the upper part of chamber B. The process is repeated as the glass is refined: it flows into C via channel b ready for working out.

The continuous tank furnace, like the continuous pot, consisted at first of three parts: the melting chamber, the refining chamber and the working chamber. The furnace was cross-fired: gas flames sweeping transversely across the surface melted the batch that had been charged into the melting chamber and the molten glass sank to the bottom of this chamber. It then passed through channels into the upper part of the refining chamber. These channels led from the bottom of the dividing wall between the melting and refining chambers, up the side of the dividing wall and out into the top of the refining chamber. This chamber was supplied with an air-cooled weir several inches below the glass surface, which brought the unrefined glass to the hot surface for refining, as in the continuous pot. The air-cooling was applied to the inner surfaces of the weir, which were not in contact with the glass. After refining, the glass passed through channels in the second dividing wall into the working end of the tank. The floor and side walls of the tank and the partition wall between the melting and refining chambers were air-cooled to give protection against the heat and the corrosive action of the glass. Accumulations of glass at points where it passed from one chamber to another could occur with consequent devitrification through standing at reduced temperature. The ascending passages were thus constricted as little as possible and the depth of the working chamber was kept very shallow for the same reason. The glass here had a depth of about ten inches. The internal dimensions of the furnace chamber itself were 24.75 x 7.5 feet and the tank had a depth of twenty inches.

With the continuous regenerative tank furnace regular operation was possible with an output per shift of about seventy-five to eighty hundredweight of glass; the production capacity was about double that of the ordinary pot furnace. The continuously maintained melting temperature saved fuel because no heat was lost during a period of cooling; there was also a saving in wages of about sixty per cent because no specially skilled worker was required for batch-charging. The old direct-fired furnace had required a melter to supervise all stages of batch-charging and melting. Pot arching and setting were also eliminated.

The new furnaces were not without their teething troubles. Refractory materials at that time could rarely withstand the demands made on them by the new furnaces. Siemens had trouble with throats becoming blocked with cold or devitrified glass and he experimented with the control of the flow of glass through the furnace with floating refractory rings in place of partition walls; he also introduced air-cooling of the outside of the tanks. He was concerned with the development of the flame inside the furnace and changed the shape of his furnaces to aid this development. A report of 1885 describes one of these designs:

> ... the gas and air inlets lead into the melting chamber at a relatively high level above the glass surface. ... The flame can develop freely and unhindered in the wide furnace and can flow through the semicircular chamber without striking or coming into direct contact with the crown, the batch, the glass surface or the partition walls. ... the tank itself has acquired an essentially different shape than it had previously. Formerly, the furnace extended lengthwise and the flames traversed it in the direction of its short dimension; now its shape approximates to that of a horseshoe, the width of which takes in the four regenerators while, in the direction of its longitudinal axis, the tank proper extends not much more than half the total breadth of the same. Instead of being spanned by a low barrel-crown as formerly, a high spherical crown now covers the melting chamber.
>
> In the direction of the long axis of the regenerators, two projections are built on the one side of the furnace to accommodate the flues rising up from the regenerators while, distributed around the semicircular side, are the working holes. ...

Modern furnaces are rectangular; the very large furnaces are cross-fired, the flames being developed across the width of the tank; many medium size furnaces are end-fired, the flame entering at the end of the furnace, burning in a horseshoe, or U-shaped, path and being exhausted through a port in the same end as the fuel enters. The revolutionary ideas of the Siemens brothers incorporated all the essential features of the modern tank furnace. Improvements in the hundred years since their inventions have been in relatively minor points of design including many advances in instrumentation and control, and the realization of the necessity for larger regenerators in order to obtain the higher temperatures made possible by improvements in refractories.

A view of part of a modern end-fired (horseshoe) furnace is shown in Figure 38. One regenerator can be seen, it is about forty feet high. The rectangular vessel at the top of the picture is the batch-hopper from which the batch is fed into the 'doghouse' at the side of the furnace. The machines are separated from the furnace hall by the partition on the left. At the bottom, on the left, are air-blowers for cooling.

Modern refractories

The interior of a large modern cross-fired tank furnace is shown in Figure 39. This is an oil-fired furnace with a melting capacity of 320 tons per day and it is viewed from the right hand side at the end where the batch is fed to the furnace. At the end at the extreme right hand side of the picture the throat can just be seen, a small rectangular opening at the level of the siege of the furnace. Five large ports are visible through which the pre-heated combustion air enters; beneath each port are two burner blocks with circular apertures through which the oil-burners inject the fuel. There are a similar set of ports and burners on the right hand side of the furnace. The vertical blocks forming the wall underneath the burner blocks are the tank blocks. These are about three and half feet high and the glass level is normally about one and a half inches below the top of these blocks. Alongside the ports are the side walls, between the side walls and the tank blocks, level with the burner blocks, are the tuckstones. The arched roof is known as the crown and above the throat the incomplete wall known as the shadow-wall separates the working end from the melting end pictured in the figure. Four thermocouple housings can be seen protruding through the siege.

Special refractory materials are chosen for the various parts of a tank furnace but until about 1930 the only types of tank blocks in general use were made of fireclay or pot clay and certain types of natural stone. The glass industry consumes only about five per cent of the total production of refractories and the industry has had a wide variety of refractories developed for other purposes to choose from for the positions such as the crown, ports, regenerator chambers, glass contact refractories, the tank blocks, the siege and special parts including the feeder forehearth and refractory parts of the feeder itself.

Fig. 38. An oil-fired regenerative furnace showing batch hopper, regenerator and cooling fan system. The regenerator is about forty feet high and the furnace is seen from the side, the forming machines are beyond the partition wall on the left hand side of the picture. The batch from the hopper melts as it moves through the furnace from right to left towards the forming machines.

Fig. 39. The interior of a modern cross-fired tank furnace. The pre-heated combustion air enters through the five large ports on the left; below are the burners through which the oil is injected. The working end is separated from the melting end by the open-work wall, the shadow wall, on the right of the picture, and the molten glass passes through the small opening, the throat, seen at the extreme right of the picture.

The story of the development of tank blocks is best followed with the phase diagram of Bowen and Greig (Figure 23) in mind, although it has to be remembered that the porosity of the block and its impurity content are most important factors.

Originally, fireclay tank blocks were made from a clay similar to that found in the Stourbridge area which contains around twenty-five per cent alumina. A natural sandstone, especially from Penshaw, County Durham, was also used in this country well into the 1920s; this sandstone contained between eighty and ninety per cent silica but it wore away evenly in the glass and was useful for the siege. At that time the glass surface temperature would be below 1400°C and the siege only about 800°C. The fireclay tank blocks lasted about a year and in the early 1920s the natural mineral sillimanite was proposed as a raw material for making tank blocks.

In 1924 when Bowen and Greig published their phase diagram and X-ray diffraction studies of silicate crystals were beginning to provide means of studying the progress of firing of tank blocks by identifying the actual crystals present, the time was ripe for a new phase of development. Sillimanite contains about forty per cent silica and fifty-five per cent alumina, it is therefore very refractory and when calcined it was used as 'grog' in the manufacture of tank blocks. In 1944 large blocks of natural sillimanite were imported from Assam and after shaping were used as tank blocks. China clay tank blocks (alumina content forty-three per cent), particularly with added alumina, were also used, but the modern glass contact refractory is nowadays almost always a block made by fusing the raw materials at very high temperatures, 1800-2000°C, in an electric arc furnace and casting into moulds. This results in blocks with large crystals and very low porosity which are highly resistant to attack by molten glass.

The first practical fusion-cast refractories were made in 1925 by G. S. Fulcher of the Corning Glass Company in the United States. He was granted a patent for making fusion-cast mullite blocks and in 1926 a second patent for the addition of ten to sixty per cent zirconia (ZrO_2) to his original alumino-silicate materials. The modern ZAC zirconia-corundum fusion-cast refractories, with the composition range thirty-two to thirty-five per cent ZrO_2, fifty-three to fifty-four per cent Al_2O_3, ten to twelve per cent SiO_2, have been developed from Fulcher's refractories and have been available commercially since 1947. They are extremely resistant to molten glass attack and are extensively used for glass contact refractories.

Modern tank furnaces have very long lives, operate at high melting temperatures, and are capable of double or treble the production rates of pre-war furnaces. For example, a tank furnace of 1940 melting soda-lime glass at about 1450°C often had a life of about one to one and a half years, whereas by 1966 the life had increased to five to six years and the melting temperature was 100°C higher than that of the 1940 furnace. The modern furnaces produce up to 200 tons of glass per square foot of melting area during the lifetime of the furnace.

The adoption of tank furnaces

The use of the regenerative tank furnace spread rapidly in the glass container industry; in 1872, out of 177 furnaces in this country forty were tank furnaces. The quality of the glass was not good and the bottle-blowers did not readily accept the introduction of the new furnaces. The Glass Bottle Makers of Yorkshire Trade Protection Society, which published codes of rules concerning the conditions of work of its members, was concerned about this new threat to their livelihoods and in a report of the Society dated 1871 the following appeal appears:

> Fellow workmen: After the 14th of January there will be 40 of us out of employment depending on your sympathy and support. We would have liked in this appeal to have given you some idea of the way we have had to labour since our master introduced the tanks, but space would fail; suffice it to say it is no uncommon thing for us to throw away ten, fifteen and as many as twenty dozen per day, and then our master is constantly complaining about seconds going to the warehouse.
>
> Fellow workmen: you that have not experienced the disadvantages connected with working at these tanks can form but little idea of the extra labour and anxiety we have to contend with and we hope that you never will.

Beatson Clark, a Yorkshire firm established in 1751, had built up a trade in high-quality specialist bottles and did not build a tank furnace until 1927. On the other hand, Kilner Brothers Glassworks, also in Yorkshire, are shown in a survey published by the Glass Bottle Makers of Yorkshire in 1872 to have had two regenerative tanks. There was apparently pressure on the firm which encouraged the installation of the new furnace, as the survey shows:

> ... Wm Lipscombe, Esq., commenced a litigation against Messrs Kilner Bros, for damaging the crops on his estate, by not consuming the smoke. ... An injunction (was granted) against Messrs Kilner Bros, and they were then under the necessity of introducing another kind of furnace, or closing their works entirely. They, however, made arrangements with Mr C. W. Siemens, Civil Engineer, London, for a licence to use his Patent Regenerative Gas Furnace.
>
> Since obtaining the returns, Messrs Kilner Bros have erected another Regenerative Gas Furnace, and also a Continuous Glass Melting Furnace.

When mechanical methods of bottle-making were introduced and when sheet glass and plate glass began to be made by continuous processes the regenerative tank furnace became a necessary part of the complete plant and its history is inseparable from the development of the machinery. Fortunately these developments ensured very great improvement in working conditions and in the last fifty years the industry has had an unrivalled record of good

Fig. 40. A modern glass container plant. The melting furnace feeds four automatic bottle making machines which produce about 700,000 bottles per day.

industrial relations. It is appropriate to conclude this section with a commentary on Figure 40 which shows a modern bottle-making installation at Beatson Clark's factory in Rotherham, Yorkshire. The view shows in the foreground two I.S. bottle-making machines and the ends of the lehrs with the transfer belts. There are two other machines and lehrs in corresponding positions at the back of the picture. Going nearer to the furnace the four feeder forehearths are visible and the external parts of the feeder mechanisms. The forehearths protrude from the working end of the furnace. At the centre, to the rear of the picture, is the melting furnace. This furnace with its four machines produces about 700,000 bottles per day. To complete this modern plant there is an automatic batch-weighing and mixing installation.

Recuperative furnaces

In the 1920s many manufacturers still used direct-fired pot furnaces and most of the pot furnaces of that period which did employ gas-firing were designed on the recuperative rather than the regenerative system. Recuperative pre-heating consists of passing the gas to be heated through a system of tubes or narrow channels around which the hot exhaust gases are circulating; by this means, heat is conducted through the dividing walls from hot to cold gas. Recuperative heating is continuous, requiring no reversal and one set of chambers only; it is much simpler than the regenerative system, but the low thermal conductivity of the recuperator walls and the tendency for these to crack and allow air and waste gases to mix are some of the problems inherent in the recuperative furnace.

These problems can be alleviated by the use of metal in place of refractory recuperators. Metal has a much higher thermal conductivity and can be better sealed to prevent air and gas mixing. The recuperator systems are carefully designed to improve heat transfer, but maximum use of waste heat from the combustion products is not possible because the metal recuperators can only operate at temperatures of about 700-900°C and waste gases must be cooled by about 1000°C before coming into contact with them. The recuperative system is now generally used for small furnaces in the hand-made industry where the glass is melted in pots.

Unit melters

Since 1951 an increasing number of small tank furnaces have been built without regeneration or recuperation. These unit melters produce high-quality glass at an economic cost, because their construction costs are low in comparison with a regenerative system and costs of repairs and maintenance are proportionately less than for the regenerative furnace. It is also possible to change very quickly between the production of different colours or compositions of batch. The unit melter is a long, narrow, direct-fired continuous glass melting furnace. It is usually oil-fired with the burners opposed to one another entering the sides near to the fining end. The exhaust gases flow from the fining zone across the melting zone towards the doghouse thus causing a heat flow counter to the batch flow resulting in an increase in the amount of batch that can be melted for a given area of furnace.

Oil-firing

In recent years oil has been increasingly used as a primary fuel in the glass industry. In 1950, oil accounted for twenty-seven per cent (150,000 tons per annum) of the primary fuel used in the United Kingdom glass industry; by 1965 this had risen to ninety per cent (650,000 tons per annum). In contrast, the glass industry's consumption of solid fuel (mainly coal and coke), fell from 400,000 tons to 80,000 tons, both expressed as oil equivalent. The consumption of gas in 1965 was equivalent to about 150,000 tons of oil. The price at that time per net therm for coal and oil was 6d ($2\frac{1}{2}$p), for gas 18d ($7\frac{1}{2}$p) and for electricity, 36d (15p). These costs are affected by economic considerations; for example, the tax on fuel-oil, and the availability of supplies. Large sources of natural gas have been discovered in northern Germany and Holland in recent years, and natural gas from the North Sea could have a similar impact upon the British glass industry provided an attractive price could be agreed. Its use in preference to oil will also be determined by other factors such as the cost of necessary piping systems and the continuing availability of suitable oil supplies. Coal and oil must be stored, whereas gas and electricity are immediately available; oil is a clean fuel and more convenient in use than coal, which produces ash and soot and is more difficult to handle. Gas and electricity are both very convenient to use and have the additional advantages of being required for other processes; gas is almost universally used for fire-polishing and electricity for motive power, and both are used for heating annealing lehrs.

Electric furnaces

Glasses are excellent insulators at room temperature but are electrolytic conductors at high temperatures. As the resistivity of the glass decreases with rising temperature, there is a rapid increase of current as the temperature of the bath rises but suitable energy regulators have been devised to control the flow of current to the glass and thus the furnace temperature. Current is passed into the glass through graphite or water-cooled molybdenum electrodes inserted through holes in the side walls or bottom blocks.

Electric heating has been used increasingly since the early 1950s to boost the output of a conventional gas- or oil-fired regenerative furnace by as much as one hundred per cent, at the same time improving the quality of the glass.

The heat introduced enters the mass of glass directly and all is utilized in heating the glass; none is lost to the furnace superstructures or waste gases. Thus the furnace can be run efficiently and at a higher temperature without damage to superstructure refractories and indeed electric furnaces have been designed in which the top of the molten glass is completely covered with unmelted batch so that only a temporary crown is required during the heating-up stage. It is also suggested that electrical heating produces convection currents which cause the cooler portions of glass near the bottom of the furnace to rise to the surface and there to come into contact with radiation produced by flame heating; as well as improving the melting efficiency, this promotes mixing of the glass, thus making it more homogeneous.

It is often more economic to practise this 'mixed melting' than to build an all-electric furnace, especially in countries where electric power is expensive. However, the capital cost of an all-electric furnace is about sixty per cent of that of a fuel-fired regenerative furnace of equal output, and repair costs are reduced as only the side-walls and throat require replacement. Efforts are being made to reduce the net heat cost per ton of glass by electric melting, by improving the thermal efficiency of the furnace and by lowering its power consumption. The initial cost of the electricity, however, often remains the major problem. In countries with abundant hydro-electric power, the practice of all-electric melting has advanced rapidly. Nearly all the glass furnaces in Switzerland, a pioneer in electric melting, operate on this system, and more than enough sheet glass is produced to supply the entire needs of the country. However, in winter hydro-electric power is in short supply and the furnaces are adapted for heating by oil.

One remarkable application of electric heating has been the provision of refining cells, into which the glass passes for electrical heating after preliminary melting in the furnace by conventional fuels. These cells are very small, usually less than six feet long by three feet wide, but because of the intensity of mixing in these cells colouring materials may be added at the point of maximum mixing and become uniformly absorbed throughout the glass; from one basic glass being melted in the tank different coloured glassware can be produced simultaneously. Colour changes can be made very quickly as the capacity of the melting chambers is small in relation to the total throughput of glass. Nevertheless, the use of such devices has not grown rapidly, one contributory reason being that in a modern mechanized glass plant production schedules are carefully planned to minimize changes on the plant.

Bibliography

1. *A textbook of Glass Technology*, F. W. Hodkin and A. Cousen, Constable & Co. Ltd., 1925.
2. *A history of technology*, (Eds.) C. Singer, E. J. Holmyard, M. R. Hall and T. I. Williams, O.U.P., 1958.

3. *De re metallica,* Georgius Agricola, (trans) H. C. and L. H. Hoover, Dover Publications, Inc., 1950.
4. *Guide du Verrier,* G. Bontemps, Paris, 1868.
5. *The development of Siemens furnaces for glass melting,* Fifth Chance Memorial Lecture, W. M. Hampton, *Chemy Ind.,* 1960, 593.
6. *Improved arrangement of furnaces which improvements are applicable in all cases where great heat is required,* F. Siemens, British Patent 2861 (1856).
7. *Improvements in furnaces,* C. W. Siemens and F. Siemens, British Patent 167 (1861).
8. *On a new construction of furnace, particularly applicable where intense heat is required,* C. W. Siemens, *Proc. Instn. mech. Engrs.,* 1857, 103.
9. *On a regenerative gas furnace as applied to glasshouses, puddling, heating, etc.,* C. W. Siemens, *Proc. Instn. mech. Engrs.,* 1862, 21.
10. *Electrical heating in glass production,* E. Meigh, *The Times review of industry,* March 1962.
11. *Concepts and economics of electric melting in the glass industry,* L. Penberthy and D. Hansen, Penberthy Electromelt Co., USA.
12. *Use of electricity in the glass industry for all electric melting, boosting and forehearth heating,* Report by M. C. Renolds, Penelectro Limited, England.
13. The French Encyclopaedia of Denis Diderot and Jean d'Alembert, 1751-65.
14. *The system $Al_2O_3.SiO_2$,* N. L. Bowen and J. W. Greig, *J. Am. Ceram. Soc.,* 1924, **7,** 238.
15. *Tank blocks for glass furnaces,* T. S. Busby, Society of Glass Technology, Sheffield, 1966.
16. *Refractories in the glass industry,* T. S. Busby, *J. Br. Ceram. Soc.,* 1966, **3,** 407.

6

Flat glass

Glass has been widely used as a glazing material since the days of the Romans but methods of producing larger sheets and more brilliant surfaces have been developed continuously and, of course, at a much increased rate in modern times. The first flat glass was of poor quality, green and bubbly, with a surface marred in making the flattened sheet. Later, glass made by the crown process had a brilliant fire-polished surface but the size of the panes was limited and, though larger sheets could be made by the hand cylinder process, the surface was again dulled by surface contact during the splitting of the cylinder and flattening to form a sheet. Thick cast plate glass, with its ground and polished surfaces, solved the problems of size and finish to a certain extent but it was very expensive to produce and little progress was made until the nineteenth century with the development of a method for grinding and polishing thin sheets of glass. During this century processes have been developed for the continuous manufacture of window glass with a natural fire-polished surface and for the continuous production of ground and polished plate glass. Recently, the new float-glass process has solved all the old problems by producing a continuous ribbon of brilliant flat glass automatically. This ribbon requires no grinding and polishing and can be formed in a great variety of colours and sizes.

Roman and Dark Age window glass

The use of glass for the glazing of windows appears to post-date its use as a material for the making of vessels by several thousand years. Nevertheless, the use of glass windows was widespread at least from early Roman times; bronze frames which were glazed with sheets of glass twenty-one inches by twenty-eight inches have been discovered at Pompeii, and the windows of the bath house in that city were of thick glass measuring in inches forty by twenty-eight by a half. Much Roman window glass was greater than one eighth of an inch in thickness, and was of a greenish-blue colour, small pieces being fitted into a more or less richly ornamented wooden frame divided into many sections. Window openings were also 'glazed' with very thin sheets of alabaster, marble or parchment, so thin that they let in light but kept out wind and rain. Window glass was probably cast as blocks: the casting process consisted simply of pouring or pressing the hot glass into flat open clay moulds, or even of pouring it out upon flat stones. Discs of glass for windows from six to eight inches in diameter first appeared in the eastern Roman

Empire in the fourth century AD, and may have been the forerunners of the Norman 'crown glass' which was made in the Middle Ages.

After the decline of the Roman Empire, window glass was still made in the West although on a reduced scale. Its chief use appears to have been for the glazing of churches, as is noted in several manuscripts of the period. An account of the miracle of St Ludger, who died in AD 809, mentions glass windows in many colours; they may have been ornamented with painting and there were probably also various ground colours. Bede states that in AD 675 French craftsmen came to glaze the church at Monkwearmouth: 'they not only did the work required but taught the English how to do it for themselves'. Before the end of the seventh century AD glass had replaced the linen and perforated boards in the windows of York Minster, but any training that the foreign glassmakers may have provided appears to have been unsuccessful, for in AD 758 the Abbot of Jarrow had to send to the Rhineland for glaziers to work on his monastery.

Stained glass and the medieval church

By the year AD 1000 conditions in Europe were becoming less warlike and church building began to flourish. A writer of that period spoke of the 'white robe of churches' which covered the land. However, the windows of these Romanesque churches were very small because the walls and pillars had to be massive and strong in order to bear the weight of structure above them. In 1137 Abbot Suger of St Denis, near Paris, started rebuilding his abbey church, and erected the first church to be designed in the Gothic style. This style, typified by the introduction of the pointed arch and later of flying buttresses which took much of the stress off the walls, enabled the builders to open up and lighten the structure and to insert jewel-like panes of glass which formed a 'wall of light', filling the church interior with glowing, insubstantial colour, and reflecting the theological idea of God as the source of perfect light. They also showed the Bible stories in vivid simplicity for those who could not read. Visitors to a strange church were instructed, in an ancient catechism, to pray to God and then to wander around the building looking at the stained glass. The popularity of stained glass spread with the new architectural style and the production of window glass began to grow at a time when vessel glass was suffering a period of decline in Europe.

Heinrich von Veldeke in his *Aeneid* of AD 1200 describes the effect made by the stained glass in a sepulchral chapel:

> 'Of garnets and of sapphires,
> Of emeralds and rubies,
> Of chrysotites and sardius,
> Topazes and beryls. . . .'

The beauty of the early stained glass is enhanced by its very imperfection and by the weathering which has taken place over the centuries. Bubbles and striations present in all old window glass cause variations in refractive index throughout the material and the finished panes, when viewed from a distance in their window setting, have a richness of colour which might be lacking in more perfect glasses. Figure 41 shows the beautiful window in Chartres Cathedral, known as 'Notre Dame de la Belle Verrière', part of which dates from the mid-twelfth century AD. The blue of the Madonna's robes and halo is paler than that of the side pieces which date from the beginning of the thirteenth century. This comparatively light colouring is characteristic of the stained glass of the twelfth century, and as the window areas grew larger in proportion to the wall, so the colours became stronger and brighter.

This great expansion in the use of coloured glass must have resulted in the medieval glassmakers performing many experiments on the production of suitable colours for the varied scenes which the windows depicted. Their colouring agents were in general those used by the ancient glassmakers as far back as Egyptian and Babylonian times, but whereas we generally have to rely upon modern chemical analysis for our understanding of these ancient glasses there is additional information about medieval glasses to be gleaned from the contemporary texts which were written about all aspects of the manufacturing industries of the time.

Although, in modern terms, only a few elements were used as colouring agents, they were obtained from many sources and their preparation was complex. Blue was one of the most important colours and was obtained from *zaffre,* an Arabic word for cobalt oxide; the material containing this oxide had to be brought from the Levant at great cost and was known as Damascus pigment. Later, cobalt ores were exported on a large scale from Saxony. The cobalt was extracted by roasting the cobalt mineral (cobalt arsenide or sulpharsenide, with various other metallic sulphides) so as to remove sulphur, arsenic and other volatile matter. Another source of cobalt used by glassmakers to give a blue colour, according to Agricola, was the residue left behind when bismuth was separated from its ore. This residue must have contained cobalt with traces of nickel, and probably of iron and copper, and the blue produced would differ from that due to pure cobalt. Combinations of *zaffre* and copper compounds (*e.g.* calcined brass) gave a sea-green.

Copper and iron compounds were used to give greens and reds. Copper was generally added as ferretto, or burnt copper, made by a recipe dating from classical times. The copper was heated with sulphur to give a black mass of copper sulphide which was then roasted until it was converted into the red ferretto. Alternatively the copper was heated with blue vitriol, or copperas, copper sulphate, the resulting material probably containing a large proportion of basic copper sulphate. Copper could also be added by the inclusion of calcined brass, or brassmakers' scales, the mixture of oxides of copper and zinc formed when brass was heated; *crocus martis,* ferric oxide, made by the same process as ferretto, but using iron instead of copper, was widely employed for yellows and browns. As with many medieval materials, the

Fig. 41. 'Notre Dame de la Belle Verrière,' mid twelfth and thirteenth centuries AD. The beautiful colours of this window, predominantly blue, were produced by the medieval craftsmen using a very restricted range of colouring agents, most of them known since ancient times.

words *crocus martis* covered many reddish compounds of iron, *e.g.* ferric acetate, nitrate or chloride made by treating iron filings with vinegar, *aqua fortis* (nitric acid) or *aqua regia* (a mixture of nitric acid with hydrochloric acid or a chloride). Opaque white glass was produced by tin oxide, and purple by manganese.

Deep red, the colour most frequently used after blue, could be produced by iron mixed with a little calcined brass but the finest reds or 'rubies' were made with copper. Copper ruby glasses could be produced by fusing glass containing copper with a small amount of tartar (potassium hydrogen tartrate) which reduced the copper to the cuprous state; the reducing atmosphere of the old furnaces also had this effect. On reheating the glass a colloidal dispersion of copper and probably of cuprous oxide was produced, giving a fine red colour. Copper ruby glass has a very deep red colour, and in order to produce a glass of the required transparency the panes of red were flashed, *i.e.* a bulb of clear glass was dipped into a pot of copper ruby glass to give a thin transparent layer of ruby glass on top of the clear glass. All medieval ruby glass was flashed, but the art of making copper ruby seems to have died out during the early seventeenth century. The French glassmaker Georges Bontemps revived its production at his factory in Choisy-le-Roi in 1826, and later brought his knowledge to Chance Brothers in England where copper ruby glass was produced on a large scale during the nineteenth century.

Gold ruby glass is generally not thought to have been made before the late-sixteenth century but a manuscript of the fifteenth century preserved in the convent of S. Salvatore, Bologna, gives recipes for dissolving gold in *aqua regia* and precipitating it by means of a solution of tin, and for obtaining a rose-coloured glass with the precipitate. A manuscript also exists in the British Museum, dated 1572, which was transcribed from an 'old copye', and gives several recipes in which gold was used as the colouring agent. Thus it seems that the seventeenth-century makers of gold ruby glass, notably Kunckel, may have had some knowledge of these earlier methods.

Details of a scene on the stained glass window, such as the folds of drapery, were applied by means of painting, and then firing, a black enamel pigment derived from iron. From the thirteenth century a second pigment in the form of a 'stain' of silver chloride or sulphide was used. When the stain was applied to the clear glass and fired, colours varying from yellow to orange were produced. The stain was later applied to blue glass to give a green colour, making possible the depiction of blue sky and green fields on the same piece of glass.

Crown glass

The Normans made crown glass, although its origins dated back to Roman times, and it was in general use until the end of the eighteenth century.

Figure 42, taken from an eighteenth-century manuscript, shows the process. Because the glass was blown it was thin, in contrast to cast plate glass. Several men and boys worked as a team in its production, the boys doing the simpler tasks. The glass was allowed to cool until it had a treacle-like consistency,

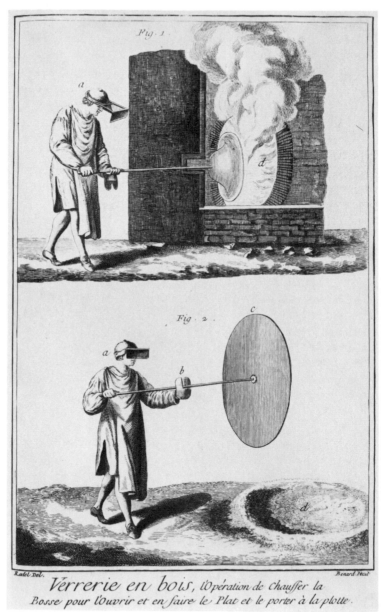

Fig. 42. Making crown glass: an eighteenth-century illustration. The glassmaker spins the opened bulb at the 'glory hole' and as it softens the spinning causes it to suddenly 'flash' or open out into a large disc.

when it could be gathered on a blowing iron. A fireclay ring was floated in the melting crucible to keep back scum which would have affected the clarity of the glass, and the glass was gathered inside the ring. As insufficient glass could be obtained by one gather, gathering was done in stages; by marvering on stone, iron or wood the gather was formed to a conical shape. An assistant blew down the pipe to expand the globe, and from time to time the glass was reheated at the furnace mouth to prevent setting. A pontil rod was attached to the globe by a knob of molten glass on the side opposite to the blowing iron which was then cracked off leaving a small jagged opening. A highly-skilled glassmaker took the globe to a reheating furnace, rotating it to keep the shape. When the globe was sufficiently heated a point was reached at which the centrifugal force caused it to 'flash' into a flat disc, attached at the centre to the pontil rod. The disc was instantly removed from the furnace and rotation was continued until it had become cooler and stiffer, when it could be cut from the rod with shears and placed in a kiln for annealing. The pieces of window glass that could be cut from the disc were small, especially if they had to be cut to form a square, but they had a brilliant fire-polish, in contrast to sheets of broad glass.

A vivid description of the crown glass process is given by Dr S. Muspratt in his book *Chemistry as applied to Arts and Manufactures* of 1854-1862. The following extract describes the flashing of the glass:

> The open projecting end of the piece, which was next to the now detached pipe, is called the nose, and gives its name to the furnace or nose-hole, where this nose is, on account of its thickness, heated almost to melting, with a view to the next operation. It is now that the glass undergoes its last and most dreadful torture in the hands of a man who, with a veil before his face, stands in front of a huge circle of flame, termed the 'flashing furnace', into which he thrusts his piece, rapidly, meanwhile, revolving his ponty. The action of heat and centrifugal force is soon visible. The nose of the piece, or hole caused by the removal of the blowing pipe, enlarges, the parts around cannot resist the tendency, the opening grows larger and larger; for a moment is caught a glimpse of a circle with a double rim; the next moment, before the eyes of the astonished spectator, is whirling a thin transparent circular plate of glass. . . . The sound of the final opening of the piece has been compared to that produced by quickly expanding a wet umbrella.

Broad glass

Theophilus was particularly interested in the making of glass plates for medieval church windows and he gave an account of broad glass making in his *Schedula Diversarum Artium.* In places Theophilus' account is not clear, but the process changed little over the centuries and Figures 43 and 44, taken from an eighteenth-century book, show how the sheet of glass was made from

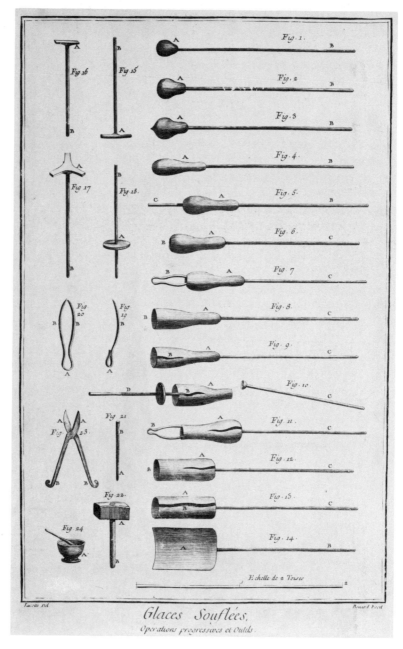

Fig. 43. Making broad glass: an eighteenth-century illustration. Figures 1 to 14 show in detail how the broad glass sheet was formed by blowing an elongated bulb of glass, opening out and widening each end in turn, and finally slitting the cylinder along its length and flattening out the sheet.

Fig. 44. Making broad glass: an eighteenth-century illustration. This illustration shows the workmen actually making the glass.

a blown cylinder. When the sheet had been opened it was placed in a 'cooling oven, moderately heated' (*i.e.* an annealing oven) with other plates and allowed to cool slowly. Broad glass could be made into larger and more useful pieces than crown glass but the finished sheet had not the fire-polish of crown glass.

Plates could also be made from a blown bulb by cutting the bulb into strips with a hot iron, reheating the strips and pressing them flat.

The forest glasshouses of England and France

From Roman times onwards France, with its vast beech forests, became a major centre of the medieval window glass industry. Up to 1500, 168 glassworks are recorded as having existed there, mostly in the fourteenth and fifteenth centuries. The situation in England was very different; a few post-Conquest glasshouses built prior to 1567 were sited in the Weald of Surrey and Sussex, and pre-Conquest furnace sites have been found near Warrington in Lancashire, at Caistor-by-Norwich (Roman) and Glastonbury (late Saxon): in all about a dozen sites are known.

There is little doubt that the medieval glassmakers in England came from Normandy, probably from the Forêt de Lyons region, east of Rouen, which had a flourishing glass industry at least from the early fourteenth century. There was an easy shipping link between Fécamp or Rouen, and Littlehampton, and the Normans could have travelled up the Arun to the woodlands of the Weald. The Abbey of St Martin at Seez in southern Normandy held land in Sussex and one furnace (1385) was on land held by this Abbey. The industry is known to have been established in Chiddingfold in the Weald by 1351, but the quality of its wares evidently never matched the famed window glass of Normandy, which was imported on a large scale. Records are very scanty until the mid-sixteenth century, but they suggest that the industry survived in a small area round Chiddingfold for at least 260 years until it moved to coal-producing areas in 1618.

The York glass painters had great difficulty in obtaining glass in the latter part of the fifteenth century and they were obliged to use English glass for white sections of the windows, suggesting that English window glass may have been in demand only when nothing else was available. In 1439 the Countess of Warwick placed an order with glazier John Pruddhe of Westminster which stated that he was to use 'no glass of England' but glass 'from beyond the seas', again suggesting that although the industry survived the glass may not have been of very high quality.

John Aubrey of Wiltshire (1626-97) stated that 'glass windows except in churches and gentlemen's houses were rare before the time of Henry the Eighth (1509-47)'; but during the sixteenth century, there was a sharp rise in the standard of living, with an increasing use of both window and vessel glass. The industry had to expand and improve in order to meet this growing demand. Coinciding with the building boom, the unsettled conditions prevailing in France during the religious wars brought about an interruption of glass supplies and thus English glasshouses were required to make good the shortage.

The Lorraine glassmakers in England

The Lorrainer glassmakers came originally from Bohemia but in the early fifteenth century they started to travel, arriving eventually in the wooded regions of the Vosges. There they settled and, having an ample fuel supply, they made good 'brode-' or broad-glass for windows by their traditional blown and slit-cylinder method, as described by Theophilus. Glassmaking secrets were handed down from father to son, preserved by strict protective oaths.

The new demands for window glass in England were met by the immigrant Lorraine glassmakers, together with families from Normandy and a few local families of the Weald. Many Lorrainers moved away from the area when they had finished their contracts spreading their improved manufacturing techniques to many parts of the country. Their glass was very costly by modern standards: the estimate for the ninth Earl of Northumberland's proposed 'New House at Petworth' in 1615, was 6d (2½p) a square foot, and the same price was charged in 1627 for glass for St Saviours', Southwark, now the Cathedral. Window glass for more ordinary buildings was perhaps 3d (about 1p) a square foot. A principal workman from a Norman family could earn 18s (90p) daily during the period of Carré's contract, whilst in 1583 a carpenter, mason and a plumber each received an average wage of 1s (5p) a day. There is no doubt, however, that many people could afford the high prices, and the industry flourished. The poorer sections of the community, meanwhile, had to do without window glass. According to Aubrey: '... before the Civil Wars (1642-49), copyholders and poor people had none in Herefordshire, Monmouthshire and Salop; it is so still.'

Nevertheless, by 1624 the descendants of the Lorrainers had established a great centre for the manufacture of window glass amidst the Northumberland and Durham coalfields. These coalfields, close to the coast from which coal could conveniently be shipped, were the first to be exploited in England. Mansell grasped the possibility of obtaining cheap coal in abundance and decided to build his window glass factories there in about 1618. By 1624 between three and four thousand cases of glass were reaching London from Newcastle every year and the industry gradually spread out along Tyneside: in 1736 Newcastle continued to provide the window glass that was 'most in use in England' and it remained the chief centre of production until the second half of the nineteenth century.

Plate glass manufacture in France

By the middle of the seventeenth century a new demand was growing, especially in France, for large, thick polished plates of high-quality glass for mirrors and coach windows. Crown glass was too thin, and the pane size too small to fulfil these requirements, although it gave a fire-polished surface to the finished glass. Broad glass was not very suitable for making into plate glass, although until the end of the seventeenth century plates were generally made by this method. Purer materials were used than for window glass and thick cylinders could be blown but only with the sacrifice of size. In England, mirrors of blown plate glass were manufactured at Lambeth in about 1670 by Italian workmen brought over by the Duke of Buckingham. In France, the beautiful panels for the Hall of Mirrors at Versailles (1678-84) were also made from blown plate glass; the mirrors are composed of small separate pieces and there is visible distortion in the panes.

A successful method for making plate glass came finally from the casting process. Some advances were made by the Venetians at the beginning of the fourteenth century, and afterwards at Nuremberg, but the plates were small. In 1676 the French minister Colbert, being impressed by the promising results obtained during the production of plate glass by casting in Normandy, placed the French royal glassworks under the control of the expert Norman glassmaker Lucas de Nehou. The inventor of the cast plate process was probably Bernard Perrot. A result of his work was the patent granted to several Frenchmen in December 1688 which gave them the monopoly of cast plate manufacture for the French home market and, later, for export. By 1691 the process had been greatly improved and satisfactory thick plates over six feet long were being manufactured.

The cast plate was initially made near Paris but in their search for fuel the glassmakers had soon to move away from this area and they sited their works in 1695 at St Gobain in Picardy where wood was plentiful and cheap. After many difficulties they were commercially successful; in 1725 they produced about 700 tons of glass and by 1760, about 1150 tons per year.

The plate glass furnace has been described in Chapter 5. The batch materials were fritted in a small furnace prior to melting and then, having been mixed with some waste glass, or cullet, were added by stages into the melting pots, which could contain 200 lb of molten glass. One layer of material was allowed to melt before the next layer was added until the pot was full, the furnace meanwhile being kept at red heat. The furnace temperature was then raised for twenty-four hours in order to fine the glass. This was the difficult stage as the furnace temperature has to be high enough to allow fining to take place, but not too high to cause the pot to fuse and drip into the glass. The pots themselves were initially uncovered, as it was unnecessary to protect the glass from sulphurous fumes and black spotting that would have occurred had the fuel been coal. From time to time the surface was skimmed with a large ladle to remove the scum.

The fined glass had then to be transported to the casting table, and this was acccomplished by various means. Molten glass was sometimes ladled straight out of the pots, but generally a square refractory container, or cuvette, was used. It was placed inside the furnace in order to warm up with the melting pots, and the refined glass was transferred from the pot to the cuvette using long ladles inserted through openings in the sides of the furnace. After the glass had stood for some hours, long hooked iron rods were slid into the furnace around and under the cuvette so that it could be withdrawn. After transfer to the casting table on a two-wheeled chariot, the surface of the molten glass was skimmed and the cuvette firmly fastened by iron hooks and chains to a pulley system whereby it could be lifted, inverted, and the contents poured onto the casting table. Figure 45 shows the casting in progress and the glass sheet being rolled out.

At a later stage the melting pot itself was lifted from the furnace, instead of using a cuvette. Of necessity the pots had to be smaller than the original ones, and the constant handling was harmful to them, but the process was more economical of molten glass and the glass itself was of better quality having been handled less.

The illustration shows how the lateral limits of the sheet were determined by movable iron rods. The iron roller ran on these rods, and the worker quickly flattened the glass to a uniform thickness; the operation had to be carried out in under a minute as the glass cooled very rapidly. When very large plates were made several men were required to operate the heavy roller. The rolled plates were then annealed for about ten days in low flat ovens, which can be seen in the background in Figure 45, after which they could be ground and polished. The removal of the plates for grinding and polishing is shown in Figure 46.

Fig. 45. Casting plate glass: an eighteenth-century illustration. The molten glass is being poured from the cuvette onto the casting table and rolled into a flat sheet which is then pushed off the table into the annealing oven on the left hand side of the picture. The two workmen at the back are pushing the carriage on which the cuvette was brought from the melting furnace.

Fig. 46. In this companion illustration to Figure 45 the workmen are removing panes of glass from the annealing oven. The panes were very large, requiring many men to lift them.

To grind the plate, it was laid upon a flat bed of very fine grained sandstone or limestone, to which it was usually cemented with plaster of paris to prevent movement. Round the bed was a protective wooden ledge. The grinding was accomplished by rubbing the abrasives on the plate to be ground with a small piece of glass; water and coarse sand were followed by finer grades of sand and finally by powdered glass. This was done by hand and was a long and tedious process until the introduction of mechanical grinding and polishing at the end of the eighteenth century. Here the glass plate used for abrasion was cemented to a wooden plank which in turn was attached to a horizontal wheel moved backwards and forwards by the operator. After the plate had been ground on both sides, it was polished using finely powdered rotton stone, a type of decomposing siliceous limestone, emery (a mixture of aluminium and iron oxides) or rouge applied by a felt roller.

Cast plate manufacture required considerable investment. The building itself was very large, a hall rather than a small glasshouse such as was customary at that period. The central melting furnace was also bigger, as were the annealing ovens to accommodate the large panes of glass, and expensive lifting, transporting, grinding and polishing equipment was also required. With fragile panes of such a size, up to one hundred and sixty by thirty inches, breakages were frequent and costly, a difficulty which brought more than one plate glassworks to the verge of bankruptcy. Labour charges were also high, as a large number of men was required to run the works.

The development had to wait until means for such investment were possible, as were provided in France by the Royal Household and the nobility. Absence of the means of investment contributed to the delay of nearly one hundred years before cast plate manufacture became established in England.

The introduction of the cast plate process in England

In 1691 a patent was granted to Robert Hookes and Christopher Dodsworth which included 'the Art of Casting Glasse and particularly Looking Glasse Plates, much larger than ever was blown in England or in any Forreigne Parts'. The patent and subsequent attempts to form a 'Company of Glass-Makers' eventually came to nothing, though in June 1692 a group of glassworks owners were advertising their wares in glowing terms: 'all sorts of exquisite Looking Glass plates, Coach-Glasses, Sash and other lustrous Glass for Windows and other Uses'. It is known that there was a fairly large trade around the turn of the century, the Vauxhall glasshouse especially making fine mirrors. The owners appear to have been unwilling to take any extended risks in expenditure on costly machinery with uncertain possibilities of financial return particularly as the manufacturers could make a comfortable profit on their existing lines of smaller size panes, and so the manufacture of cast plate fell into complete disuse.

By the second half of the eighteenth century the situation had completely altered. The demand for fine large plates arose in England just as it had in France years earlier. By 1773 it was estimated that the French factories were supplying the English with between £60,000 and £100,000 worth of plate glass per year, and it was alleged that illegal trading companies were formed to smuggle it in at a cheaper rate. People were now eager to put money into such an obviously expanding enterprise and accordingly, in the early 1770s, steps were taken to recommence manufacture in England. The sponsors of the projected British Cast Plate Glass Manufacturers Company sought incorporation as a joint-stock company for twenty-one years by a private Act of Parliament. This allowed them to raise a joint-stock of £40,000 from shareholders and a further £20,000 with the permission of three-quarters of these holders. Incorporation by private Act of Parliament was a new development towards the end of the eighteenth century and it entailed limited liability which, whilst helping the manufacturers in carrying on a risky enterprise, was considered by a large section of the community as being contrary to the public interest. A Commons Committee was set up and the members summoned a former St Gobain worker, Philip Besnard, as a technical expert. He was of the opinion that cast plate manufacture in England was feasible because all raw materials, excepting barilla for soda, could be obtained locally, and if coal were used as the fuel the fuel costs would be less in England than in France. The Act of Incorporation was finally passed in April 1773.

Low fuel costs were probably the decisive factor in the siting of the new glassworks, which were built at Ravenhead, near St Helens, Lancashire, in 1773. One of the sponsors had considerable interests in the coalmines of the region. The Cheshire saltworks and the Liverpool export trade were the main outlets for coal, but he realized that local furnace industries, such as copper smelting and glassmaking, would create profitable and easy markets. Accordingly he offered exceptionally favourable terms for his high-grade coal which may explain why the industry was sited at Ravenhead rather than on Tyneside.

The first manager of the works was probably a French-trained worker, Jean Baptist Francis Graux de la Bruyère, and most of the experienced workmen were also French. Poor management, high excise duties on the manufacturing process and a high breakage rate contributed to the lack of progress of the new company but the French workers were unable to use coal for firing the furnaces and it was not until 1792, when Robert Sherbourne became the new manager, that covered, or caped, pots were successfully introduced to prevent spotting of the molten glass. Eventually the company was reorganized, competition from St Gobain was greatly reduced by war, and cast plate manufacture in England became commercially successful.

The casting process was at first the same as that used in France, but improvements were made as time went on. The first casting tables at Ravenhead were of copper supported on solid stone, as were those used in France at that time. When hot glass was poured onto these tables the copper could crack easily, incurring much expense, labour and loss of time in replacing them. In about 1843 large iron plates became available and, after the fear that iron would discolour the glass had been overcome, iron tables mounted on castors which could be wheeled to any annealing oven replaced the stationary copper tables. Improvements were also made in the polishing process when in 1789 a steam engine, built by Boulton and Watt, was used to provide the power.

During the first half of the nineteenth century the manufacture of plate glass in Great Britain was carried on by about six firms of which three, including the Ravenhead works, were in the St Helens area. By the late 1860s these three factories were said to produce two-thirds of Britain's output of plate glass, but the Ravenhead works were now experiencing serious difficulties and in 1868 they were taken over by the London and Manchester Plate Glass Company. During this period other firms were showing an interest in plate glass manufacture, especially Pilkington Brothers of St Helens. This firm had originally been founded in 1826 to make window glass when it was known as the St Helens Crown Glass Company. From the first the company had been associated with a wide variety of successful enterprises; the original partners included William Pilkington, a surgeon's son who also managed a prosperous wine and spirit business in St Helens, and Peter Greenall, the manager of the St Helens Brewery and a leading promoter of the first St Helens Building Society. Throughout the nineteenth century the Pilkingtons showed considerable foresight in anticipating consumer demands and in acquiring interests in the operations of their competitors. In 1873 they decided to build a plate glass works at Cowley Hill, St Helens to supply the growing demand for plate glass; the plant was completed in about three years and by 1878 plate glass was being made on a scale comparable with that at Ravenhead. Windle Pilkington, the nephew of William Pilkington, devised many technical improvements such as a new form of movable crane for carrying the pots to the casting table and an improved annealing kiln which provided a more uniform heat to the glass plates. These innovations enabled them to make large profits on plate glass at a time when British exports to the USA were being drastically cut because of the growth of the American plate glass industry. Belgium, the main supplier of plate glass to the USA, tried to gain new markets in Great Britain and thus competition for home markets also increased. Most British plate glass companies, already making very low profits, collapsed and in 1901 Pilkingtons were able to acquire the Ravenhead works which they used to supplement the capacity of Cowley Hill. Their production continued to grow and by the early twentieth century they were the sole manufacturers of plate glass in Great Britain.

The cylinder process for the manufacture of window glass

The cylinder process for the manufacture of window glass was developed from the old broad glass process in Lorraine and the German states. A larger cylinder was blown than was usual for the manufacture of broad glass, by swinging the globe of glass in a deep trench. As the globe elongated a cylindrical shape was maintained by blowing into it and a finished cylinder fifty to seventy inches long and twelve to twenty inches in diameter was formed. The cylinder was allowed to cool, slit longitudinally with a hot iron or a diamond cutter and then reheated in a flattening oven (usually called a lear) where it was opened out to form a flat sheet on lagres, on the floor of the lear.

By the 1830s the cylinder process had been adopted on a very large scale in the United States and on the Continent, but most of the window glass in England was still made by the crown process and very little cylinder glass was imported owing to the high import duties. The English were accustomed to small bright panes in their windows, and as the process of crown glassmaking was brought to a state of great perfection in England, it survived there for longer than on the Continent, only declining as other methods of producing sheet glass grew in importance during the 1860s. Attempts were made prior to 1832 by Isaac Cookson and Company to introduce the manufacture of cylinder glass to England, but they were unsuccessful, the glass failing to compete with even the poorest quality crown glass. In 1832 Lucas Chance of Chance Brothers, Birmingham and Georges Bontemps embarked upon the project of making cylinder glass at Chance's works in Spon Lane, employing foreign workmen who were skilled in the process.

Chance and Bontemps determined to make 'all the good glass we possibly can, and find out the best market for the bad'. They were able to fulfil their second aim by exporting poor quality glass to British possessions which were protected against foreign competition: the grade of crown glass just above the 'coarse' was already known as 'Irish'. In contrast to the situation on the Continent, there was a large demand for good glass in England and the cylinder process was modified in several ways to supply this demand. For example the cylinders were blown wider and glass of better surface quality was obtained. Improvements were also made in the selection of clean, pure batch materials, the sand being washed and sodium carbonate or sulphate being used instead of kelp. Careful trials of different batch mixtures were made and eventually it was reported that 'our light colour has evidently given our glass a character in the market which it never had before.'

Chance's patent plate glass

The quality of the glass was improved by these measures, but the lack of surface brilliance still remained as the sheets came into contact with other surfaces during the flattening operations. Thick blown plate could be ground and polished, but when thinner, larger cylinders were made, the glass could not be ground in this way as sheets of less than one quarter of an inch in thickness either broke under the strain of grinding or, owing to their uneven surfaces, wore through in places into holes. In 1838, James Timmins Chance, the nephew of Lucas Chance, patented a method which solved the problem completely. During grinding and polishing the sheets were made to remain perfectly flat by making them stick by suction to slates covered with leather or other suitable material soaked in water. The backing for the sheets was quite firm, but it was not so rigid as to impose a strain on the glass during processing. Thus the risk of breakage was greatly reduced and the glass was ground uniformly over its surface.

The patent plate process was an immediate success and by late 1845 production was about 10,000 feet a week, the glass being in great demand for coach windows, ornamental mirrors and coverings for pictures. It had a less brilliant surface than crown glass but larger panes could be made, limited only by the size of the original blown cylinder and this made it very useful where large areas of transparent glass were required.

Glass for the Crystal Palace

At the time that they were developing patent plate, Chance Brothers were also expanding their production of unpolished cylinder sheet glass. In 1850 they were awarded the contract for the glazing of the Crystal Palace in preparation for the Great International Exhibition of 1851. This contract required the production in a few months of 200 tons of sheet glass over and above their normal output and Robert Chance, the son of Lucas Chance, was hurried off to Lyons to engage thirty skilled blowers to undertake the extra work. Existing employees were also called upon to work night and day and one of Chance's abandoned works was re-opened. Production was continually increased until, in January 1851, sixty-three thousand panes of sixteen ounce glass were being turned out in a fortnight. Figure 47 shows the blowers at work forming the huge cylinders. By the end of the month the work was completed and nearly one million square feet of sheet glass were supplied to the Palace.

Hartley's patent rolled plate glass

During the 1840s James Hartley of Messrs Hartley, Sunderland, developed a method for making sheets of thin cast plate which was essentially the same as the French process of casting thick plate glass by filling a cuvette with molten glass from the melting pot and transferring it to the casting table; in Hartley's process the glass was ladled straight from the melting pot onto the casting table and rolled, thus eliminating the intermediate stage of the refining cuvette. The patent was granted in 1847 and the cast glass was known as patent plate, not to be confused with Chance's patent plate which was polished cylinder glass. The plates were often made from coloured glass and could be impressed with patterns; for most purposes they were used unpolished, as skylights, roof glass and for church windows. During annealing

Fig. 47. Chance Brothers supplied nearly one million square feet of sheet glass for the Crystal Palace. Work continued night and day for many weeks and 200 tons of glass over and above the normal output were produced.

they could be stacked in the lehr like crown and sheet glass, and the usual high cost of annealing plate glass was thus greatly reduced.

Patent rolled plate could be cast in any required size and with a minimum thickness of one eighth of an inch. Hartleys were able to tender for the Great Exhibition building of 1851 at a price only slightly greater than that of Chance Brothers, whose tender for cylinder sheet glass was finally accepted. The size of the panes of patent rolled plate was sixty-two by twenty-one inches, instead of forty-nine by ten inches, would have greatly reduced the amount of framing, but it was considered that there was too much risk involved in employing such heavy panes, which had not yet been tested extensively for glazing. However, larger sheets of thin glass, one eighth to one quarter of an inch thick, could be made by this method than by any other in use at the time, and for this reason manufacture of patent rolled plate, also known as rolled plate or rough-rolled, was soon taken up by other firms on a large scale. The market for the new glass continued to grow, and by the late 1870s large quantities of coloured rolled cathedral glass for churches were being produced.

The Lubbers cylinder-blowing machine

The hand-cylinder and rolled plate processes for the production of window glass were still in use in 1903 when the American Window Glass Company introduced machines which blew cylinders automatically. Several unsuccessful attempts had been made to mechanize the process during the nineteenth century when in about 1896 John H. Lubbers, an American window glass

Fig. 48. The Lubbers cylinder machine. The cylinder is being drawn from the pot of molten glass by a flanged metal disc or 'bait'; the diameter of the cylinder is kept constant by compressed air blown down the blow pipe attached to the centre of the bait. When the pot is nearly empty the cylinder breaks away from the bottom of the pot and is then lowered onto supports.

flattener, started to experiment with cylinder machines. His patents were acquired by the American Window Glass Company in 1903 and by 1905 the process was providing serious competition for the hand-cylinder method. It was introduced into Britain in 1910 and was still in use in 1933 at Pilkington Brothers Ltd.

Fig. 49. The Lubbers cylinder machine. The supports into which the cylinder of glass is lowered for splitting, flattening and annealing.

In the Lubbers process (Figures 48/9) sufficient molten glass for one cylinder was ladled from the furnace into a heated pot above which was the drawing apparatus. A blowpipe, with a flanged metal disc or bait fixed to its end, was lowered into the glass which solidified around the flange. The bait was then slowly raised between the guiding shafts and at the same time compressed air was blown down the pipe until the glass was blown out to the required diameter; this diameter remained constant as the cylinder was drawn upwards from the glass. The thickness of the cylinder and its diameter were controlled by the speed of drawing and the air-pressure. When the pot was nearly empty the drawing speed suddenly increased and the cylinder broke away from the bottom of the pot; it could then be lowered onto a support and cut into several sections, each of which was split, reheated, flattened and annealed. The pot which had contained the glass for the cylinder was inverted so that its underside, also in the form of a pot, was ready to receive another charge of glass for the next cylinder. Surplus glass was melted from the first pot and it was heated ready for using again.

The early Lubbers machines had two faults which were corrected as time went on. When compressed air was introduced through the blowpipe and came into contact with the hot glass, it expanded rapidly causing undulations in the glass. Also, the cylinder would often crack away from the flange of the bait during drawing and fall to the ground because the glass and the metal of the collar had different coefficients of expansion. A problem which was never entirely solved was the removal of ripple marks in the glass surface caused when the cylinder was flattened out in the lehr. However, the window glass made by the Lubbers cylinder process was far superior to that made by the hand method. Skilled workmen were able to blow cylinders measuring five or six feet in length and twelve or fourteen inches in diameter, but cylinders made by the Lubbers machine around 1917-20 were at least five times as long and twenty-one to twenty-four inches in diameter. Thus the hand manufacture of window glass was gradually abandoned as the Lubbers machine was widely adopted, but this in turn was eventually displaced by mechanical sheet-drawing machines.

Mechanical drawing of sheet glass: the Fourcault and Colburn machines

The idea of producing flat sheets of glass by drawing direct from the furnace was considered as early as 1857, when William Clark of Pittsburg was granted an English patent for a sheet-drawing process. His idea was to dip a bait into the molten glass to which the glass could attach itself and then slowly to withdraw the bait; a continuous sheet was to be produced by altering the speed of withdrawal according to the rate of setting of the glass.

Clark's process failed, as did all attempts by other workers for the next fifty years, because they were unable to prevent the sheet from narrowing as it moved upwards. The sheet edge attached to the bait was of full width, but as the molten glass was hot enough to flow for some time after it had been drawn the vertical edges approached one another.

In an attempt to overcome this problem, several workers tried to draw the sheet downwards by allowing the glass to flow through a slit, but as the sheet became heavier, the unset portion was drawn out and finally pulled away completely from the glass in the furnace. Attempts were also made to support the glass between two rollers as it emerged from the slit but the finished sheet lost much of its transparency.

Emile Fourcault of Belgium first achieved success in the drawing of flat sheets. His method, like Clark's, consisted of drawing a sheet vertically from the furnace but narrowing was avoided by forcing the molten glass to rise under hydrostatic pressure through a slit depressed below the surface of the glass. The slit was located in a fireclay float known as the debiteuse which is shown in Figure 50. This float could be placed at any desired depth below the glass surface, thus altering the speed at which the glass was forced through the slit. Normally the trough would fill with glass but as the glass emerged from the slit it was taken up by a bait and drawn upwards as a sheet. The trough constantly forced molten glass through the slit and if the ribbon was drawn away at the same rate there was very little force upon it which would tend to stretch it, thus the width of the sheet remained constant. Figure 50c shows the process in schematic form. The glass was solidified by two water-cooled

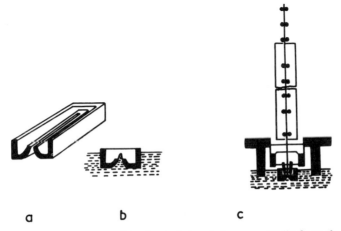

a   b   c

Fig. 50. The Fourcault process. The sheet of glass is drawn vertically from the furnace and narrowing of the sheet is prevented by forcing the molten glass to rise under hydrostatic pressure through a slit in a fireclay float (a) depressed below the surface of the glass (b). As the glass emerges from the slit it is drawn upwards by a bait and solidified by two water-cooled tubes placed against the sides of the slit. The sheet moves upwards continuously (c) guided by asbestos rollers.

tubes placed against the sides of the slit, and annealing took place as the glass rose slowly through the drawing chamber, guided by non-chilling asbestos rollers which did not mar the fire-polished surface finish.

The Fourcault machine, after many technical setbacks, began to operate commercially in 1913. During the same period Irving W. Colburn in the USA was also working on a sheet-drawing process. He started his research in 1900 and in a few years patented a machine which was not a commercial success until it was developed in conjunction with Messrs Libbey and Owens in Toledo, Ohio.

The Colburn process was similar to that of Fourcault in that a sheet was drawn up by means of a bait but instead of using a float a pair of small channelled rollers was placed just above the glass surface at each edge of the sheet; the rollers gripped and pulled it as it emerged and so maintained a uniform width. When the sheet had risen for a few feet it was softened by gas jets and bent over a horizontal roller to pass into a horizontal annealing lehr.

The Fourcault and the Colburn processes both have disadvantages. In the Fourcault process the float wears away, and causes characteristic lines on the glass known as music lines, whilst in the Colburn process the glass surface is marked when the sheet is bent over the roller. As with all sheet-drawing processes the temperatures necessary for drawing the glass are below the temperature at which a soda-lime-silica glass containing approximately equal amounts of lime and soda will devitrify. In the Fourcault process this caused a gradual build-up of crystals along the slit of the fireclay float which spoiled the glass surface. The difficulty was overcome by substituting magnesia or alumina for part of the calcium oxide; this lowered the temperature at which crystallization would occur, below the temperature of the drawing chamber.

A third flat-drawing process has been developed by the Pittsburgh Plate Glass Company which avoids these inherent defects. The sheet is pulled directly from the molten glass as in the Colburn process and is then drawn vertically upwards as in the Fourcault process. To overcome the tendency of the sheet of glass to narrow, or waist, when drawn in this manner, the glass is gripped by a pair of water-cooled, knurled rollers placed just above the level of glass in the furnace. These rollers cool the sheet and give it sufficient rigidity to prevent waisting as it rises upwards.

In 1950 all three processes were still in use; approximately seventy-two per cent of the world's production of sheet glass was made on Fourcault machines, twenty per cent on Colburn machines and the remainder by the Pittsburgh type processes. Roughly one tank furnace would feed four machines, giving a daily output of around 250 tons so that the glass for the Crystal Palace would absorb approximately two days' production.

Rolled sheet glass: the Mason-Conqueror machine

In 1884, Frederick Mason and John Conqueror patented a machine for the production of rolled sheet glass. In their first design the glass was poured down an inclined plane and passed between a pair of iron rollers to form a continuous sheet which could then travel into the annealing kiln. In practice this idea did not work, and the sheets had to be transported to the kiln on carriages. A machine rolling process was developed by Chance Brothers, and after many experiments it was successful in 1887. It was later possible to make figured sheets with a pattern impressed by the rollers on one side. In 1890 Edward Chance added a second pair of rollers which impressed a pattern on one side of the sheet after it had been formed by the first pair of rollers. Although rolled sheet was not transparent the patterned surface gave a translucency which made the sheets very popular.

The 'double-roll' machine was also used for the manufacture of wired glass, glass sheet in which wire-mesh is embedded to give strength, and patents for this process was taken out by Chance Brothers in 1905. Differences in the thermal expansion of wire and glass tended to cause cracking of the plate on cooling until an iron-nickel alloy of similar expansion coefficient to that of the glass was introduced. Care also had to be taken to prevent trapping air bubbles between the glass and the metal which weakened the finished sheet. In the modern process for the manufacture of wired glass the wire netting is laid on a thin ribbon of glass and covered with a ribbon of molten glass, the whole then being pressed through a pair of rollers to form a single sheet.

The mechanization of the cast plate process

Bessemer's rolled-plate experiments

In 1846 Henry Bessemer, famous for his work in the steel industry, patented a method of producing sheets of glass by passing the molten glass between rollers, slicing the continuous sheet with cutters attached to the upper roller, and allowing the sheets to slide down an inclined plane straight into the lehr. A necessary condition for this process was the use of a continuous melting furnace; Bessemer proposed a rectangular basin heated from above containing the glass which could stream out from a slit in the bottom of one side onto the rollers. He later wrote in his autobiography:

> Up to this period the fusion of glass in large crucibles was universal, and the reverberatory furnace which I erected at Baxter House for this purpose was the first in which glass was made on an open hearth, and the parent of all the bottle furnaces in which the fusion of glass is carried on in open tanks.

Unfortunately Bessemer's experiments, carried out largely in co-operation with Chance Brothers, were not commercially successful. The process itself was sound but Bessemer seems to have been over-impatient and his remedies for technical difficulties, which constantly cropped up, were liable to involve the experimenters in much expensive and time-wasting work. His conception of a continuous tank furnace providing molten glass for rolling as plate glass was not developed as a practical process until 1923 when Pilkington Brothers initiated the continuous plate glass process. By this time, the widespread adoption of the regenerative tank furnace made the imaginative idea a practical proposition.

The Bicheroux process

Until the end of the First World War plate glass was made by the traditional process which, with slight modification, had been practised since 1688 but the glass was melted in pots containing about one ton of molten glass which yielded a plate of about three hundred square feet. The pots were removed by a crane and carried to a water-cooled steel casting table. The contents of the pot were poured on the table in front of a roller which traversed the table. For one quarter of an inch plate glass the plate was cast about $\frac{15}{32}$ inches thick. After annealing, these plates were mounted in plaster of paris on circular tables about thirty-six feet in diameter for grinding and polishing.

In the Bicheroux process which was introduced in the early 1920s, glass was still melted in large pots but was poured onto the casting table through a pair of rollers (Figure 51), as suggested by Bessemer. Because the charge of glass remained in a large mass until it met the rollers to be formed into a plate, instead of being poured out onto the table in front of the roller, it was possible to produce much flatter plates which required much less grinding.

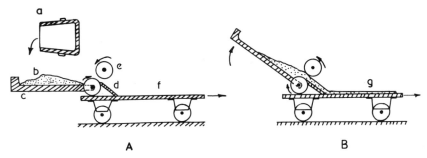

A   **B**

Fig. 51. The Bicheroux process. A. the casting table before rolling; B. the casting table raised to permit rolling; a. casting pot; b. mass of glass; c. casting table; d. inclined plane; e. rollers, rotating in directions shown by arrows; f. transportation table; g. rolled glass sheet.

The Ford process for the continuous manufacture of plate glass

The advent of the motor car caused a rapidly increasing demand for polished plate glass. The consumption of glass was so high that Henry Ford, of the Ford Motor Company, Detroit, decided to go into the glass business himself and the company eventually developed a very successful process for the continuous manufacture of plate glass.

Ford placed in charge of his plate glass plant C. W. Avery, who, when the experiments commenced in 1919, knew nothing about glass manufacture at all. After several trial meltings of glass a furnace was built with a pit below it into which a ribbon of glass could be poured but at this stage the project showed definite signs of failure. However, Ford continued to supply extensive financial support and Avery persisted with his research. By the end of 1921 all the mechanical problems had been solved and a method devised of making the glass flow from the furnace through a pair of rollers continuously; but the glass quality was very poor. At this stage, Pilkington Brothers who by this time had considerable experience of tank furnaces took an interest in the experiments. Agreement was reached for the installation of the Ford continuous plate-rolling process at Cowley Hill in 1922. Pilkington's experience with tank furnaces combined with Ford's mechanical expertise to make the process a commercial success.

Fords were mainly concerned with a narrow ribbon of glass for car windows, but their flow process could also produce a wide ribbon. The glass was melted in a tank furnace and fed between a pair of rollers, after which it passed straight into the annealing lehr. This was the first process in which the glass moved continuously from tank to lehr, contrasting greatly with the Bicheroux process in which glass was poured intermittently from pots and where the formed sheets were cut and carried to the lehr.

The continuous grinding and polishing machine

The glass made by the flow process was translucent rather than transparent because it came into contact with the rollers and in order to produce polished plate glass continuous grinding and polishing was needed. The finishing machines themselves had been driven mechanically as early as 1789, when steam was used to provide the motive power in an attempt to reduce costs and make the process more rapid. Steam power was eventually replaced by electric power and by the 1920s plate glass was ground upon a large, circular table using successively finer and finer grades of sand with water until the process was complete. The plates of glass were cemented to the table for finishing and the table was rotated under the grinding discs, large circular discs embedded with many small iron blocks which swept the whole table evenly while it rotated. After grinding, the table was rolled to the polishing machine where, by a similar process, the surface was polished using water and rouge.

At the same time that the Ford flow process was being developed at Cowley Hill, Pilkingtons were making experiments on the continuous grinding and polishing process. The large rough sheets were fixed on cast iron tables which passed under a series of grinding heads and subsequently under a set of polishing heads. An experimental machine was in operation by 1920 but the process was not operated on a commercial scale until 1923.

The twin grinding and polishing machine

The continuous grinding and polishing machine only treated one side at a time of large sheets. In order to make the process truly continuous a twin grinder and polisher was needed so that the ribbon could pass from the lehr continuously through the grinding and polishing process without being cut. A twin grinding process was developed by Pilkingtons in the early 1930s and came into service in 1935 at their Doncaster works. The polishing part of the twin grinder and polisher (Figure 52) was never completely successful and usually the ribbon was cut and the large sheets polished separately, but on a continuous machine as in the first continuous grinding and polishing process.

The flow process used in conjunction with the twin grinder was adopted by the world's main manufacturers of plate glass for most of their production; other methods, being slow and very uneconomical, fell into disuse. Plate glass production has always required a greater capital outlay than is necessary for other types of glass manufacture, and the introduction of the continuous process intensified the trend towards fewer units supplying a larger share of the total market demands.

Fig. 52. The twin grinder and polisher. The continuous ribbon of rough plate glass from the annealing lehr passes between twin grinding heads which grind both sides of the glass at once, after which it is polished.

The float-glass process

Glass produced by the continuous plate process is very flat, but requires grinding and polishing. Sheet-drawn glass on the other hand, has a brilliant fire-polished finish but the sheets show distortion caused by small differences in viscosity which affect the sheet thickness during the upward pulling motion of the forming machine. The ideal form of flat glass would therefore combine the flatness and freeness from distortion of plate glass with the natural finish and cheapness of sheet glass, and this ideal was achieved in the recent development of the float-glass process by Pilkington Brothers.

The aim of the float process was to reheat the newly-formed ribbon of glass and allow it to cool without touching a solid surface. In October 1952, Alistair Pilkington began experiments using liquid metal to support the glass ribbon as it emerged from the rollers. He says that:

> The basic idea is a continuous ribbon of glass moving out of the melting furnace and floating along on the surface of molten metal at a strictly controlled temperature. Because the glass has never touched anything while it is soft except a liquid the surface is unspoiled—it is the natural surface which melted glass forms for itself when it cools from liquid to solid. Because the surface of the liquid metal is dead flat, the glass is dead flat too. Natural forces of weight and surface tension bring it to an absolutely uniform thickness.

The float process is shown diagrammatically in Figure 53. The batch is melted in an oil-fired regenerative furnace and the glass emerges as a ribbon to float on the surface of molten tin at a carefully controlled temperature. The furnace atmosphere is controlled to prevent oxidation of the tin and heat applied from above melts the glass sufficiently so that it can conform to the flat surface of the molten tin. After sufficient cooling the plate can be fed onto rollers in the annealing lehr without affecting the surface finish and the annealed glass is cut into the required lengths.

This process took seven years of intensive work before it became a commercial proposition in 1959. A pilot plant was running in 1954 which provided a continuous flow of glass but many problems had to be solved, including the control of the float bath atmosphere, glass flow and the formation of the ribbon.

When these problems had been solved the float process was found to have many advantages. The brilliant ribbon of glass emerging from the float bath has few surface flaws and suffers little distortion. The speed of ribbon formation appears only to be limited by the melting capacity of the furnace and speeds in excess of fourteen metres per minute have been attained, the glass moving continuously through a horizontal annealing lehr to final inspection and packaging. Float-glass can be produced in a wide variety of widths and thicknesses at a cost very much less than the equivalent cost for the twin grinding process.

The first glass made by the float process had a natural thickness of about six millimetres determined by the forces of gravitation and surface tension acting upon the ribbon. Within five years of its commercial introduction methods had been developed to make thinner and thicker ribbons, the former by stretching the glass in a gentle and controlled way and the latter by

162

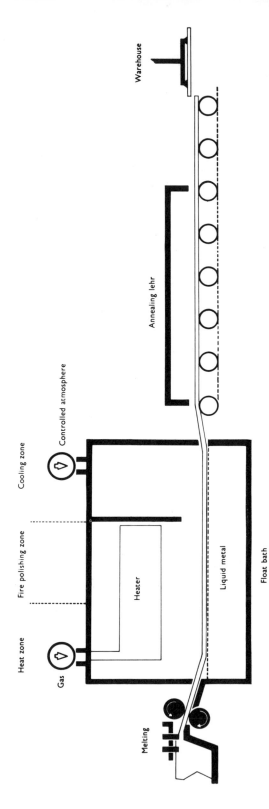

Fig. 53. The float glass process. Molten glass flows from the tank in a continuous ribbon to float on the surface of liquid tin at a carefully controlled temperature. The flat brilliant surfaces of the glass do not require grinding and polishing and the ribbon passes continuously through the annealing lehr to be cut up into the required lengths in the warehouse.

allowing the glass to build up to a certain extent within the float bath. By these methods thicknesses of between three and fifteen millimetres were produced.

The electro-float process

The electro-float process, announced in 1967, is a method whereby the surface of the clear ribbon glass can be modified during manufacture without shutting down the production line or requiring a separate unit for manufacture. As the glass passes through the float bath chamber an electric current is made to flow from an electrode above the glass through to the tin. By a rapid process of ion replacement between metal and glass, metallic particles such as copper can be implanted in the glass to controlled levels and densities.

Varying combinations of metals and temperatures yield glasses with different heat and light transmission characteristics which can be used for such purposes as reducing glare and cutting down heat transmission. Such glasses are most useful for the glazing of buildings containing large window areas, and can also be used to advantage in cars and aircraft.

Bibliography

1. *Medieval stained glass*, H. Hutter, (trans.) A. Shenfield, Methuen and Co. Ltd., 1964.
2. *Some comments on the medieval glass industry in France and England*, G. H. Kenyon, *J. Soc. Glass Technol.*, 1959, **43**, 17N.
3. *The glass industry of the Weald*, G. H. Kenyon, Leicester University Press, 1967.
4. *A history of industrial chemistry*, F. Sherwood Taylor, Heinemann, 1957.
5. *The Bicheroux process of making plate glass*, *Glass Ind.*, 1927, 8(9), 207.
6. Review Lecture, *The float glass process*, L. A. B. Pilkington, *Proc. R. Soc. Lond. A*, 1969, **314**, 1.
7. *Glass in the modern world*, F. J. Terence Maloney, Aldus, 1967.
8. *Pilkington Brothers and the glass industry*, T. C. Barker, George Allen and Unwin, 1960.
9. *Revolution in glassmaking*, W. C. Scoville, Harvard University Press, 1948.
10. *A textbook of Glass Technology*, F. W. Hodkin and A. Cousen, Constable and Co. Ltd., 1925.
11. *A history of Chance Brothers and Company*, J. F. Chance, privately printed by Spottiswoode, Ballantyne and Co. Ltd., 1919.
12. *The development of coloured glass in England*, H. J. Powell, *J. Soc. Glass Technol.*, 1922, **6**, 249.

7

Containers

Glass has been formed into hollow objects for at least 3500 years, but the art of shaping glass on the end of a blowpipe does not seem to have been invented until around the beginning of the Christian era, and glass-forming machinery which was developed at the end of the nineteenth century was a commercial success almost entirely in the present century.

Vessels and fragments of vessels dating from 1500-1450 BC have been found on western Asiatic sites and in Egypt, all apparently made by the slow and laborious method of wrapping glass around a core. At some time during the first century BC the solid metal rod which was used to hold the core was replaced by a longer hollow tube and glass-blowing was invented.

Glass-blowing in the Roman Empire

Most of the early glass-blowers worked in Alexandria and Sidon producing at first free-blown vessels which imitated those in alabaster or pottery, but later using two- or three-section moulds of wood or clay. Thus the means of mass production were available to supply the new markets opened up by the Roman conquests. Glass became utilitarian and inexpensive and new containers were developed which could hold a wider range of products. Food could be contained in jars with wide, flat bottoms which stood upright as opposed to many Egyptian containers which, formed round a core, were narrow-based and required a stand to hold them; wines, oils and medicines could also be conveniently stored and transported. Glassmakers were proud of their wares and often marked their moulds with their trade names, for example, Jason, Artas and Ennion.

The introduction of blowing led to the rapid development of narrow-neck vessels in which oil, wine and beer could be stored; for food the mouth of the container was made larger. The new range of products accentuated the problems of sealing the vessels and during the Roman era sealing techniques included the use of cloth or cloth soaked in wax or oils, fibres soaked in oil, and grease or wax gobs. Linen cloth treated with oils or resins has a very long life and was used as a 'tie-on' cover from earliest times for sealing vessels which contained both liquid and dry materials. It is uncertain whether cork was used as a stopper by the Romans as the writings which refer to cork may refer either to bark in general or specifically to cork, the bark of the cork tree.

After the fall of the Roman Empire the use of glass declined and styles became simpler; the bottle in plate 13 between pages 36 and 37, made in the

164

seventh century A D, may be compared with the Roman bottle of the second to third centuries A D shown in plate 11. The more simple styles do not reflect a general loss of technological skills but rather the different tastes of the Teutonic patrons. The gradual revival of the art of glassmaking in the Frankish Empire does not appear to have greatly affected the manufacture of containers and only small bottles, such as those used by doctors and alchemists, appear in significant quantities throughout the Middle Ages.

Manufacture of bottles by hand

The methods of manufacture of glass bottles altered very little for nearly two thousand years. At the beginning of the nineteenth century bottles were blown off-hand without a mould or in an open single piece mould which only shaped the bottom half or body of the bottle. The upper or shoulder parts together with the finish, the top of the neck, were shaped while still soft by applying a tool as the bottle was rotated. The word finish is still used, even though this is the first part of the bottle to be formed on all glass container machines.

The demand for all types of bottles increased rapidly during the early nineteenth century. Hinged moulds which could be opened and shut were introduced, permitting the production of more complicated body shapes. The outside body, shoulders and neck of the bottle could be formed together but there still remained the task of applying the final form to the finish. In one method, the finish glass was reheated and the contour was rolled on, in another the top was cracked off and ground down, and for very cheap bottles the finish might be just cracked off. The hinged moulds were at first opened and shut by boys, but later moulds closed by foot-operated levers known as mechanical boys were introduced. The pontil rod was replaced by a 'snap' clip which could hold the container without marking it, unlike the pontil rod which had to be broken from the underside of the bottle leaving a jagged surface.

The development of the wine bottle

The use of glass bottles was greatly stimulated by the extensive use of cork as a stopper from about 1650. Until the beginning of the seventeenth century nearly all bottles were made of earthenware, metal, wood or leather. Early stoppers were made of wax or resin mixtures, but cork is mentioned in English literature in the early 1500s as a material for bottle stoppers. The first corks were sealed to the bottles by being dipped in a wax compound or in oil, but by the early seventeenth century conical corks were tied down with thread and later with wire to the string-ring, a rim around the neck of the bottle just below the mouth.

HG–12

Fig. 54. The evolution of the wine bottle. The first wine bottles were little more than blown bulbs of glass but as techniques improved they assumed a cylindrical shape and, by 1750, a bottle very similar to the modern port bottle was in use.

The use of the cork stopper coincided with the Restoration in 1660 of Charles II. The great increase in wine drinking at this time led to the almost universal adoption in England of the corked glass bottle. Tightly fitting corks could be used after the introduction of the cork screw at some time prior to 1686. This may seem to be an unimportant innovation but it led to horizontal storage, the maturing of wine in bottles, and the possibility of producing sparkling wines by the *methode champenoise.* Corks were certainly more suitable than the stoppers of 'tow soaked in oil' which were previously used for champagne, though these continued in use until the nineteenth century for medicine containers.

The shape of the wine bottle continued to change throughout the seventeenth and eighteenth centuries. The earliest wine containers had been bulbous with a round base, pale green in colour and very light in weight. They were kept upright by special metal table stands or wanded by encasing in osier basketwork woven at the glassworks; some Italian wines are still bottled in a similar way. Bottles manufactured during the first half of the seventeenth century are generally described as shaft-and-globe, being nearly globular in body with a long slender neck. This shape was easy to blow but difficult to balance and in the third quarter of the seventeenth century the body became squatter, the neck became shorter and the base was flattened to provide more stable support. The kick-up, or basal concavity, was increased in size; many reasons for the kick-up have been offered but whatever the real explanation it certainly improved the stability. Although the appearance of these bottles was attractive they were difficult to store and a taller, narrower bottle with straight, slanting sides was developed after about 1715. By 1750, the slanting sides had become vertical and a cylindrical bottle, very similar to the modern port bottle, was in use. Figure 54 shows the development of the wine bottle.

The dates of manufacture of many of these bottles are known from the seals they carry; the earliest known seal dates from 1652. Glassworks of the seventeenth and eighteenth centuries made these embossed seals which were affixed to the bottles sold to clubs, inns and the wealthier members of the general public. Some carried the date and the initials or crest of the buyer and were sealed to all bottles intended for the client. Wines and spirits were stored by the inns and the customers had their bottles filled from these supplies. The volume of trade may be judged from the fact that by 1695, 240,000 dozens of bottles were made in England every year, chiefly for table use but also, especially in private houses, for storage. Incidentally, in 1969 nearly 489 million dozens of containers were made in Great Britain, in other words about one hundred containers for every man, woman and child.

The beer bottle

Beer and wine have been known since the beginning of recorded history, but until the seventeenth century beer was stored in wooden casks and either drawn straight from the barrel or carried in ceramic or leather containers. The discovery that beer can be preserved in bottles is attributed to Dr Alexander Nowell, Dean of St Paul's from 1560 to 1602. He was a very keen fisherman and one day accidentally left his bottle of ale on the river bank. Finding it several days later he opened it and discovered 'no bottle, but a gun, so great was the sound at the opening thereof'. Secondary fermentation had occurred and the quality of the ale had improved. As the seventeenth century progressed, home-brewed ale was stored in bottles, which were made either of earthenware or glass and the shape of the glass beer bottle evolved in a similar way to that of the wine bottle. Housewives of the early seventeenth century were advised to bottle ale in round, narrow-mouthed bottles 'stopping them close with a cork ... fast tied in with strong Pack thread, for fear of rising out, and taking vent, which is the utter spoil of the ale'. This method of closure was also used for wines of the same period.

The English bottle-makers of the later seventeenth century not only supplied this domestic market but produced beer and wine bottles for export. The earlier 'tonnage and poundage' statutes taxed bottles imported into the country, and not until 1660 do glass bottles appear in the 'Export' list; presumably prior to this period the English glassmakers were not making sufficient bottles even to supply the home market. The making of glass bottles was greatly encouraged towards the end of the seventeenth century by the demand in London for West Country liquors, such as beer, cider and even the famous Bristol Hotwell's Water, the demand for which was so great that it gave considerable impetus to the bottle industries of Bristol and Gloucester.

During the late eighteenth and nineteenth centuries, prior to automatic bottle-making, a great variety of bottle shapes were introduced, some of them poorly adapted to the blowing process. Bottles intended merely for containers were black or very dark green through excess of iron and other impurities in the raw materials. In 1831, in evidence given at an excise enquiry, it was stated that the materials employed in common bottle manufacture were sand, soap-makers' waste (for soda), lime, common clay and ground bricks.

The first pasteurized beer, following on the work of Louis Pasteur on fermentation, was produced in Copenhagen in 1870. Pasteur recognized that the agents causing fermentation were micro-organisms and he was able to give valuable advice to wine and beer manufacturers on the diseases attacking their products. Pasteurized beer did not rapidly go stale and could be bottled on a commercial scale. The cork inlaid crown cap which had been patented in the USA in 1892, gradually replaced the cork and by 1912 it was in almost universal use in that country for sealing bottles containing drinks under pressure. Until very recently it remained the most popular method for sealing beer and soft drinks but now one-trip bottles with a screw top and a modified twist-off closure are coming into increasing use.

Bottles for soft drinks

The drinking of natural spa water became a popular habit amongst the English during the eighteenth century, though special waters had been known prior to this period for their curative properties. They were bottled in earthenware containers in order to increase their sale, but the public at large probably found them rather unpalatable. Attempts to produce substitutes for natural water date back to 1560 when it was recognized that 'gas' was present in the water. Cavendish succeeded in producing the gas, carbon dioxide, by chemical reaction in 1766 and Joseph Priestly, in the 1770s, succeeded in developing the first practical method of making artificial mineral water; commercial production of soda-water began at Manchester in 1777.

The earthenware bottles used by the early mineral water manufacturers were unsatisfactory as they were permeable at high gaseous pressures, thus heavy glass bottles soon came into use. The soda-water syphon was invented in 1815 and soda-water rapidly became popular as a retail bottled product. At the Great Exhibition of 1851 over one million bottles of flavoured mineral water or 'pop' as it was then known, were consumed. In 1814 William Hamilton patented an egg-shaped bottle for artificial mineral water. Because of its shape it had a much greater resistance to high internal pressures than the bottles in general use at the time, and as it had to be stored on its side the cork was kept moist, thus preventing leakage of carbon dioxide through a dry stopper. This bottle was widely used in England after 1840 until the end of the century, when it was replaced by the flat-egg bottle, which was easier to fill and could be stored on its side or on its flat narrow base. It was adapted to take the crown cap in about 1903 and remained popular until the late 1920s. Figure 55 illustrates a selection of nineteenth-century soft drink bottles.

The availability of rubber resulted in several inventions concerned with bottle closures. Gutta-percha was brought to Europe in 1843 and the vulcanization of caoutchouc, or rubber, was invented around 1842 by Charles Goodyear in the USA. The invention was not patented in England until 1844 and Thomas Hancock in London produced a similar material in 1843. The internal screw stopper and the swing stopper were invented in the 1870s and the famous Codd bottle was patented by Hiram Codd of Camberwell in 1875. It contained a glass marble which was kept pressed against a rubber ring in the neck of the bottle by the internal gas pressure thus forming an excellent hermetic seal which was released when the marble was forced downwards. Its period of greatest popularity in Britain was from 1890 to 1914, though it continued to be used in this country until the 1930s and is still manufactured in the Far East. The crown cap gradually replaced most of the earlier stoppers after the First World War, when bottles with this type of closure were sent in large numbers to American forces serving overseas.

Fig. 55. Nineteenth-century soft drink bottles. In the foreground is the Hamilton bottle; in a semi-circle from left to right, a tall narrow Seltzer bottle, a 'flat-egg' bottle, a Reliance Patent ball-stoppered bottle, and another Seltzer bottle.

Milk bottles

The first milk to be marketed in Great Britain in glass bottles was introduced in the 1880s by George Barham, who was then the managing director of the Express Dairy Company. Sterilized filtered milk in swing-stoppered bottles was introduced in 1894; until after the First World War, however, milk continued to be sold from churns often pushed around the streets in hand carts. Conditions were still unhygienic and formaldehyde or borax were often added to combat tuberculosis, diphtheria and typhoid fever. Pasteurized bottled milk, sealed at first with cardboard discs, and much later with aluminium tops, was delivered to the house soon after the First World War and it gradually spread through the country. Pasteurization is a process of partial sterilization during which the bacteria are killed by heating the milk to around 72° C for fifteen to twenty seconds.

Containers for foodstuffs

Glass containers have been used since Roman times as receptacles for food, but their employment as agents in the preservation of food followed developments in the fields of agriculture, chemistry and nutritional studies which took place in the seventeenth and eighteenth centuries. Perishable food could not be preserved and the diet of all classes at this time was severely limited. The quality of staple food such as bread rose during the seventeenth century, but between 1530 and 1640 wages in England rose more slowly than food prices and most people were badly fed during this period. They lived on beans, some salted meat, bread, fish, cheese and a little bacon or game. Townspeople, who were in a minority before the Industrial Revolution, fared better and ate reasonable amounts of butter, meat, white bread and fruit and an increasing quantity of vegetables and fresh fish. Nevertheless, by the end of the winter all classes were reduced to a monotonous unbalanced diet of beer, salted port, dried beans and barley bread, and many people contracted scurvy.

The improvement of the diet was not aided during the seventeenth century by the current ideas on nutrition. They were still governed by the Aristotelian concept of the four elements, which was the basis of the 'humoral theory' of nutrition. For example, vegetables were distrusted on the grounds that they engendered wind and melancholy. Foodstuffs were submitted to destructive distillation in order to analyse them; the 'watery' or 'oily' and 'saline' products were then examined and many unfounded conclusions drawn. Deficiency diseases were not recognized as such and indeed were virtually untreated before 1750, except in situations where they were particularly rife such as on long sea voyages. Drake is known to have sailed round the world well provided with wood sorrel which is rich in vitamin C.

In the 1730s, Townshend's development of rootcrop cultivation made possible the winter feeding of cattle and the continuous availability of fresh meat. By 1780, Irish housewives were preserving various foods by cooking them in pint glass jars and sealing on the lids with carpenters' glue but the idea of treatment of food at high temperatures was developed by Nicholas Appert, who in 1810 won Napoleon's prize for a method of carrying food to the army in the field. In his book, *The art of preserving,* published in 1810, he described his theory that heat would prevent fruits, meats, fish and vegetables from deterioration. He chose glass as the material for his containers because air could not penetrate through it and his jars were sealed by long corks firmly driven into their mouths.

The preserving industry spread from France to England after the Napoleonic Wars and the first American preserving company opened in Boston in 1819. The early preserves were sometimes stored in glass containers but as hand-made containers had irregular sealing surfaces it was very difficult to form airtight seals and throughout most of the nineteenth century glass was not widely used for foodstuffs. In 1858 the highly successful Mason jar was patented by an American, John Landis Mason, but it was not a commercial proposition until the final development of automatic machinery for container manufacture at the beginning of the twentieth century. The Mason jar, a wide-mouthed screw jar with the top edge ground which was used with a metal or glass cap, was the standard container used by American housewives from 1860 to 1910 and machine-made Mason jars are still used for home preserving. The quality of the ordinary jam jar was also improved over the years and in 1940 was so good that simple tin plate lids sealed with rubber bands and metal clips could be used for preserving.

The introduction of automatic machinery was quickly followed by suitable closure methods for foodstuffs. Phenolic resins, developed to a large extent for radio components during the First World War were adapted to make moulded bottle caps during the 1920s and urea moulding compounds, which could be made in white and light pastel colours, were first used in the early 1930s. The roll-on closure, the tear-off aluminium cap and the pry-off cap were all introduced in the 1920s. The past twenty-five years have seen the introduction of the soft plastic snap-cap and stopper, the lectraseal (for products such as instant coffee) and the tear-off crown for which no opener is required, as well as the related design of closure tops and container bottoms to facilitate stacking.

Glass for medicine, perfumes and cosmetics

The apothecaries of the Middle Ages were much concerned with the use of the 'urynall', or, more correctly, a medical inspection bottle, described in 1688 as 'a clear and thyn glasse bottle, with a long neck and round body; it is used by doctors, Apothecaries and such as follow phisick, to put in the water of diseased bodyes for them to looke at and to give their judgement of the distemper'. This practice was eventually banned as a diagnostic method owing to its abuse by unqualified persons.

Another extensive use for glass bottles arose during the late eighteenth and early nineteenth centuries when patent medicines became popular. One could buy anything from hair restorer to stomach medicine and the vendors often travelled from town to town, or displayed their wares at country fairs, prudently removing themselves when the populace discovered that the product did not live up to the highly coloured claims on the label.

Today, medicine bottles are produced in a wide variety of shapes and sizes; for example, rounds, ovals, flats and panels. The Winchesters are round bottles ranging in capacity from a half to forty fluid ounces, although an eighty fluid ounce size known commonly but incorrectly as a Winchester quart is also made. Ampoules designed to contain small doses of liquids or powders are manufactured automatically from thin-walled glass tubing and sealed off after filling; the end can easily be broken off when the contents are required.

Small glass bottles were used during the seventeenth century for perfumes and smelling salts but the introduction of pressing during the nineteenth century made it possible to produce elaborate shapes very cheaply, and this trend was continued with the introduction of fully automatic processes. Today both fully automatic and semi-automatic methods are used, the latter for more specialized products and many colours and finishes are available.

Mechanization of the container industry during the mid-nineteenth century

During the early nineteenth century the demand for all types of glass grew rapidly and the first section of the glass industry to meet this demand was the pressed glass industry. The early attempts of the container manufacturers were not so successful and though many machines were constructed they did not reach a practical stage until the late nineteenth century.

The first patent for a bottle-making machine was taken out by Alexander Mein of Glasgow in 1859 followed closely by one due to C. G. W. and J. Kilner of Yorkshire for a glass bottle-blowing machine (1860), and a more advanced design by James Bowron of Stockton-on-Tees (1861). In 1876, A. R. Weber took out a patent in the USA for a machine similar to that of James Bowron but in which pressure on the glass was used to form a more perfect mouth. All these experimental machines were aimed at forming a narrow-mouthed article. In 1875, the first attempt was made in the USA to devise a mechanism for making wide-mouth containers. The idea was to take a sufficient amount of glass, the gob, from the furnace, shape it to an initial rough jar shape, the parison or blank, by pressing and then to obtain the finished container from the blank by a blowing process. A modern version of these operations is shown schematically in Figure 66. This first press-and-blow machine was designed by James S. and Thomas B. Atterbury of Pittsburgh with the object of making jugs, but it never seems to have been put to practical use.

The Arbogast patent for the press-and-blow process

Philip Arbogast of Pittsburgh was the first man to design a machine incorporating the three essential steps found in all successful bottle machines to the present day. These steps are the formation of the top of the neck of the article, followed by the formation of the parison and finally by blowing to the desired shape. In 1882 Arbogast was granted a patent for his invention, which he claimed as an 'improvement in the manufacture of glassware, consisting in pressing the mouth or neck to finished form with a dependent mass of glass, then withdrawing the plunger, then removing the article from the press mould and finally inserting it in a separate mould and blowing to form the body'.

Arbogast was unable to develop a practical machine and he assigned his patent rights to the D. C. Ripley Company of Pittsburgh in 1885, but the union, which effectively controlled Ripley's factory, fixed high wage rates and small outputs, making the use of his machines uneconomic. The Ripley Company was taken over by the United States Glass Company which then granted manufacturing licences to non-union firms and Vaseline jars were made under licence on Arbogast press-and-blow machines by the Enterprise Glass Company in 1893.

Some firms working with their own versions of press-and-blow machines ran into difficulties over patent rights. A machine made in 1886 by Charles E. Blue for the Atlas Glass Company caused legal disputes with the United States Glass Company and Blue was forced to take out a licence under the Arbogast patent. His machine, unlike Arbogast's machine, did not require the parison to be transferred into another blow mould. After pressing, the blank mould fell out of the blow mould and the parison could then be blown as it hung in the blow mould. Blue's machines were widely used in the USA at the beginning of the twentieth century for the manufacture of fruit jars and other wide-mouth containers and were put into use in 1900 by Messrs Kilner Brothers of Yorkshire.

The press-and-blow process for the production of jars was also developed during the same period in England. In 1886, J. R. Windmill devised a practical process and his method was soon put into commercial operation. Dan Rylands of the Rylands Glass Company, Stairfoot, Barnsley, also developed a successful process in the early nineties; both the Windmill and the Rylands machines were used for many years. Thus the introduction of commercially successful press-and-blow machines in England and America occurred at about the same period and Windmill and Rylands were in conflict for a time with Blue over patent rights.

The Ashley bottle machines

The press-and-blow process could not be used for narrow-mouthed bottles because the opening was too small to allow the plunger to be inserted and then withdrawn. The first machine to make narrow-mouthed bottles was patented in 1886 by Josiah C. Arnall and Howard M. Ashley. In 1866 Arnall was the Postmaster at Ferrybridge, Yorkshire, and in the course of his duties he visited the local bottle works and noted the way in which the bottles were painstakingly made by hand. He conceived an idea for blowing bottles by machinery but when he submitted his plan to a prominent south Yorkshire bottle manufacturer, it was turned down as being too crude and far too revolutionary for development. After a lapse of nearly twenty years he had a second opportunity to discuss his ideas, this time with Howard Ashley, who was the manager of an iron foundry in Ferrybridge and who had never been employed in a glassworks. As a result of the discussion, Ashley built an experimental machine in which molten glass was poured into an inverted mould fitted with a plug to form the inside of the neck of the bottle and with a sliding plunger fitting loosely into the body of the mould from its open end. A charge of glass was dropped into the mould and was pressed about the neck-forming plug by the plunger. Compressed air was then admitted through the formed neck, blowing up the bottle and raising the plunger.

Ashley was encouraged by his results to try to develop a practical machine. He replaced the single mould by three distinct moulds, one for forming the neck of the bottle (the ring mould), the second for giving an initial form to the bottle (the parison mould) and the third for final shaping of the bottle (the blow mould). The three moulds were incorporated into all Ashley's later machines and in all successful bottle machines since. Their functions corresponded to the stages of making a bottle by hand, except that in hand-making the gather is worked and blown up into the body of the bottle whilst still on the gathering iron and the finish can only be formed after the completed body has been cracked off the iron. Ashley built his machine at Ferrybridge and it was developed at Armley in Leeds in 1886 and later at Castleford. A machine of this type is shown in Figure 56 and the principles of its operation are illustrated in Figure 57, which is a modern version of the original Ashley process. The charge of glass was placed in the parison mould and the plunger pressed upwards into the glass to form the neck. The parison mould was opened and removed by hand and the head carrying the parison by the neck rotated until the parison was hanging vertically from the neck. The blow mould halves were brought into position by the action of a foot pedal and closed. The bottle was then blown to fill the mould.

Although bottles were made successfully on this machine it needed several men to operate it. According to a contemporary report, one skilled gatherer with about six unskilled helpers could feed two of these machines and produce about one gross of soda-water bottles per hour. A team of five hand workers of whom three would be described as skilled could produce bottles at the same rate. Ashley therefore designed a rotary machine which was built in Sheffield with four parison moulds, four corresponding ring moulds and one blow mould, the ring moulds and blow moulds being hinged to the same arms that carried the blowing heads. The bottle was formed in the same way as on the single unit machine, but the mould table was rotated mechanically step by step at a pre-determined rate. Glass was placed in the inverted parison

Fig. 56. The Ashley plank and pillar bottle blowing machine, 1889. The Ashley machine, developed by Howard M. Ashley of Yorkshire, was the first practical machine to make narrow-mouthed bottles by a semi-automatic process.

mould and the neck was formed by a plunger during a one-step rotation. During the next step the plunger was withdrawn, air was puffed into the blank and the parison mould and blow head were inverted. The parison mould halves opened automatically leaving the parison hanging freely from the ring mould. The first blowing operation was carried out in a blow mould mounted on a separate stand which was brought up and closed round the blank after which the bottle was released during the third stage of rotation.

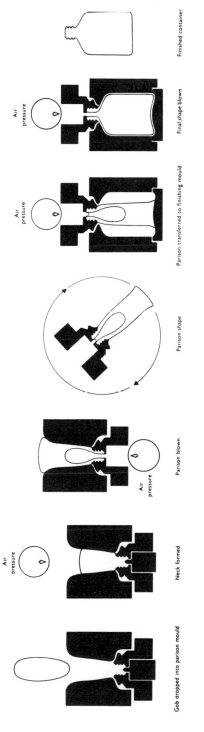

Fig. 57. Semi-automatic blowing of narrow-mouthed containers. The gob is gathered by hand on an iron and the correct amount of glass is dropped into a preliminary mould where the neck is formed by compressed air. The rough shape, the parison, is then transferred to the finishing mould in which the final shape is blown.

The blow head and its attached parts were again mechanically inverted and the parison mould closed ready to receive another charge of glass.

In later versions of this machine the four units were replaced by three and the opening and closing of the parison moulds were accomplished by a piston operated by compressed air. These machines required two men for their operation and were each capable of turning out eighteen dozen bottles per hour.

The patent rights for the Ashley machines were at first bought by Messrs Sykes, MacVay, and the Codd Bottle Company of Castleford in 1887 but were taken over in 1888 by the newly-formed Ashley (Machine Made) Bottle Company. Shares were quickly oversubscribed on the publication of a glowing prospectus by the promoters and the Leeds *Mercury* of December 18th 1887 described the process at Messrs Sykes, MacVay and Company, in the following terms:

> Sykes, MacVay and Company's New Process at Castleford. Another familiar landmark is going. The glass bottle trade is in process of being melted down into new 'parisons' without blowpipes and blowers and instead of five men being necessary to evolve an imperial receptacle for beer or aerated water, it almost looks as if five innocently occupied adults might discover pastime in watching the conjoined labour of a machine and a youth in placing bottles at the service of good liquor as fast as they can be counted. . . . Never since the days of the Pharaohs has anything so clever in glass-making been devised, nor anything so simple. . . . It has remained for a Yorkshireman Mr H. M. Ashley, of Ferrybridge to revolutionize the trade . . . the merest novice standing by could see that the invention is equal to its task and is a grand simplication of a much sub-divided, and imperfect operation as carried out in its present form. . . . In the works of Sykes and Company, sixty (furnace) holes yield 420 gross per day (by hand production), but the machine production will yield 4800 gross per day.

However, possibly owing to the gross inefficiency of the management and strong union hostility, the company did not prosper. The machines themselves functioned extremely well and Messrs Bagley, and Cannington Shaw, who purchased the Ashley machines at the final auction in November 1894 succeeded in making good bottles after adding improvements of their own.

The development of the semi-automatic bottle machine: 1890-1918

Although Ashley was the first to develop a commercial blow-and-blow process for bottle-making there was considerable interest in the development of this type of semi-automatic machine in France and Germany as well as in England. Various machines were built on the Continent during the 1890's without much success, with the notable exception of one developed during the period 1894-1903 by Claude Boucher for the Verreries de Cognac. This was similar to the Ashley machine, but a controlled blast of compressed air was directed at the parison in order to stiffen it and to prevent it from elongating too much during transfer to the blow mould, thus ensuring more even glass distribution. Progress in both France and Germany lagged behind that in England but in about 1905, A. Schiller of Germany constructed a blow-and-blow machine employing suction to form the finish, which came onto the market in 1906. After various modifications and improvements the Schiller machine became established as one of the leading semi-automatic machines. Schiller also developed a press-and-blow machine during the same period and it is estimated that from 1906 to 1932, 1150 of the machines of the two types were supplied to the glass industry.

In England, mechanization of the industry proceeded rapidly during the opening years of the twentieth century and by 1907 at least fourteen firms were operating bottle machines, mainly in Yorkshire and Lancashire. Joshua Horne, who had built many machines for the Ashley Bottle Company, played an important part in the development. In 1901 he took out a patent in his own name; the machines which he built under this patent came into widespread use in the north of England and were also supplied to France, Germany and the USA. By 1914-15 over one hundred Ashley-type machines were in use in the USA where several improved versions had been developed. All these machines were semi-automatic, that is, they required to be fed by a gatherer who fed the glass to the mould by allowing it to drop from the iron into the parison mould; when sufficient had fallen the hot glass was cut by shears from the remainder on the rod.

In order for fully automatic bottle-blowing to be realized some means of gathering and feeding had to be invented. This problem was solved in two ways, both of which are still in use on various machines, although the first successful method, the suction feed, is used less frequently. In the suction method just enough glass is sucked up by vacuum into a suction mould, whilst in the gravity feed method the glass flows out through a hole in the floor of an extension to the furnace and is cut off or sheared when sufficient to make a bottle, commonly called a gob, has flowed through. Both of these methods needed large reservoirs of molten glass and could only be developed successfully when the regenerative tank furnace came into use.

The Owens suction machines

The development of automatic bottle machines which eventually replaced both hand and semi-automatic processes dates from the end of the nineteenth century and is associated with the name of M. J. Owens.

Michael Joseph Owens, born in 1859 in West Virgina, was sent to work at the age of ten years in the flint glass factory of Hobbs, Brockunier and

Company, at Wheeling, West Virginia. Conditions of child labour at that time were exceedingly bad and the hardships experienced by Owens must have influenced his later work in removing many of these abuses. In 1880, 23.4 per cent of the workers in the bottle industry of the USA were under sixteen years of age and they worked in shifts, the night shift lasting from 5.30 pm to 3.30 am. Children of nine years of age worked under appalling conditions in the glass factories and were often permanently scarred by severe burns. Working under these adverse conditions, and receiving little formal schooling, Owens made great efforts to teach himself and was an active member of a local debating society and a branch leader in the American Flint Glass Workers Union. In 1888, Owens joined the W. L. Libbey and Son Glass Company (later known as the Libbey Glass Company) as a second-ranking blower in a small group making glass shades for oil lamps. After only three months he became foreman of the blowing department and two years later, through his exceptional talents, superintendent of the factory.

Owens was now able to devote time to the problem of blowing and forming glass by mechanical means and in 1894 he obtained two patents for mechanically operating paste moulds, which he assigned to the Libbey Glass Company for the manufacture of light bulbs. Paste moulds are metal moulds coated with carbon and then sprinkled with oil or water before each blowing, the bulbs are rotated during blowing and marks due to the joins of the mould are eliminated from the finished product. This venture led to the formation, in 1895, of the Toledo Glass Company to which Owens and his sponsor, Edward Drummond Libbey, assigned their remaining rights in the paste mould machine and agreed to pass over all patents that they might obtain during the next seventeen years. The Toledo Glass Company then changed its interest from tumbler and lamp-chimney machines to developing completely automatic bottle machines. Libbey supported Owens throughout his experimental work and provided the necessary finance.

The first Owens machine worked rather like a large bicycle pump or hand spray gun. Figure 58 shows this apparatus in use. The 'gun' sucked up a charge from the furnace into a two-part mould, the upper part of which formed the neck and lip around a plunger and the lower part of which was designed to enclose the requisite gather of glass to form the parison. After sucking up the glass by the withdrawal of the piston rod, the glass was cut off by a sliding knife which formed the bottom plate of the two-part mould. Then the entire gather was carried to a table and the parison, suspended by its neck from the ring mould, was transferred to the finishing mould where the final shape was blown by pushing in the piston rod.

Although this hand 'gun', costing about fifty dollars, was not a practical proposition (see Figure 59), it embodied some of the most important principles, such as gathering by suction, utilizing the full-sized separate mould for forming the neck of the bottle and alternating the use of gathering and blowing moulds, found in later machines. Owens did not originate any one of

Fig. 58. Owens experimental apparatus No. 1. The device gathered glass by suction and the neck and lip of the bottle were formed by a plunger. The parison, suspended by its neck from the neck ring mould, was transferred to the finishing mould, seen at the left of the picture, where the final shape was blown by pushing in the piston rod.

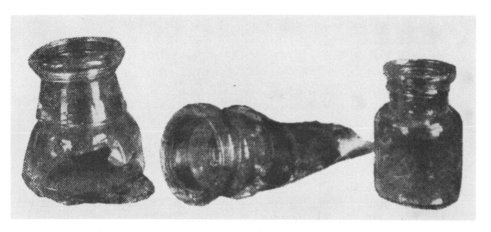

Fig. 59. These twisted shapes were the first attempts to make bottles by the Owens vacuum method.

HG—13

these principles but he and his helpers succeeded in combining the separate features to form a workable automatic mechanism.

Owens modified this machine by mounting the gun on a column provided with wheels so that it could be moved to and from its own small glass tank. The third machine was intended to be an automatic rotary machine having a series of arms carrying gathering and finishing moulds, but a test model with only one complete gathering and finishing mould soon showed that further gathering at a fixed point was impossible. Excess glass gathered from the furnace dropped back and, being cooler and therefore more viscous than the glass in the tank, it formed a cooler inhomogeneous area around the gathering point. To avoid this difficulty a special furnace was built which fed glass to a pot continuously revolving at such a speed that the place where the gather had occurred was reheated before any fresh gather was made near it. A machine with six arms, each carrying a gathering and a finishing mould was so successful that the Owens Bottle Machine Company was set up in 1903; the Owens machine by 1907 had reduced the wages cost of making a gross of pint beer bottles from $1.50 to 10 cents.

In 1905 the Owens British Bottle Machine Company was formed in order to promote the machines in Europe. This company built a small works at Trafford Park, Manchester in 1906, and by 1907 Owens machines had been installed and were successfully producing bottles. A licence had previously been granted to the Rheinahr Glassworks of the Apollinaris Company in Germany, solely for the manufacture of mineral water bottles. The European manufacturers were very worried by the threat of continuing competition from the Manchester works and from the Apollinaris Company and in 1907 all the patent rights of the Owens machines which had been assigned to the new Owens European Bottle Machine Company, together with specified arrangements for further help and advice and the rights to improvements, were sold for 12,000,000 German marks (£600,000) to the Europaischer Verband der Flaschenfabriken Gesellschaft, the EV. British manufacturers, organized as the British Association of Glass Bottle Manufacturers Limited, put up 2,400,000 marks of the purchasing price, and the use of Owens machines quickly spread throughout the country.

The Apollinaris Company did remarkably well out of the deal because they had initially bought one-fifth of the shares of the Owens European Bottle Machine Company. Thus, when the agreement was made with the EV their share of the 12,000,000 marks gave them a gross profit of about £92,480 after deducting their original share of the purchase price. They had then only to find about £20,000 in order to pay for their modern factory using the latest automatic machines.

Further models were developed by Owens in the light of manufacturing experience. His fifth machine had a series of six fixed arms carrying the gathering moulds. The whole machine was raised and lowered by a system of counterweights as the gathering mould dipped into the glass while the machine was revolving. The intermittent stopping of rotation of the machine and revolving tank, which was a feature of his fourth model, was eliminated and an increase in machine speed and therefore in bottle output became possible. Dissatisfied with this machine, Owens developed a sixth, more efficient, model known as the A type and by the end of 1911 one hundred and three of these machines had been installed with a capacity of over four million gross of bottles per year. In 1911 he adopted dipping-heads in the ten-arm machines. It was thus unnecessary for the whole machine to be raised and lowered as each individual head consisted of a complete bottle-forming unit and each dipped in turn into the glass. The number of arms on the machines were increased and also the number of heads on each arm. When equipped with two, three or four cavity moulds, they were capable of tremendous outputs, limited finally by the necessity for a large auxiliary furnace for the gathering pots with consequent increased fuel consumption. Nevertheless, Owens machines were adopted by well over half the American bottle industry and were in widespread use for over fifty years. Figures 60 and 61 show in diagrammatic form the operation of the Owens machine and the turning cycle as the machine rotates.

In comparison with hand manufacture the labour costs of these machines were very much reduced. Where bottles were made by hand a team, or shop, of three skilled men and three 'boys' or helpers could make between fifteen and twenty gross of pint beer bottles in every eight and one-half hour working day. This team of six was known as the 'American' team, as opposed to the usual arrangement in Britain where a team of five was employed for the hand manufacture of many types of bottle. The American method gave a much higher productivity rate, about twice that of the British method, and thus the American production figures probably represent the maximum rate of hand production at the time of the introduction of the Owens machine. One man and five boys working twelve hours a day could operate two machines simultaneously and during a working day could make at least as many bottles as could be made by hand by fifteen men and as many helpers. Thus the machine saved man-hours and it also reduced the number of expensive skilled workmen.

Automatic feeders

The Owens machine brought about the first completely mechanized production of bottles. Following this work the production scheme evolved of automatic batch weighing and mixing and delivery of batch to a tank furnace, which supplies glass to bottle-blowing machines, from which the finished bottles are taken and stacked mechanically into the continuous annealing lehr. Only the inspection of the finished product has yet to be completely mechanized although considerable progress has already been made.

HG—13*

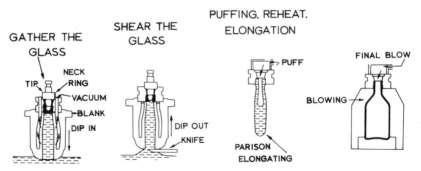

Fig. 60. The operation of the Owens machine: how the bottle is formed (see also figure below).

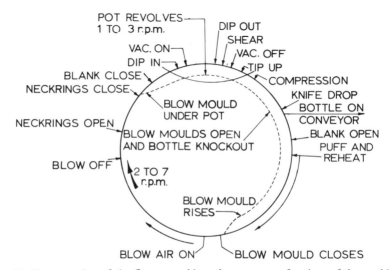

Fig. 61. The operation of the Owens machine: the sequence of actions of the machine.

It is essential in this process that the glass is removed from the furnace in amounts just sufficient to make a bottle. Owens accomplished this by the suction method, a process which is still employed in the Westlake machine and in some Roirant bottle machines, but the device by which this step is now generally accomplished is known in the industry as a gob feeder, a mechanical device which frees the blowing machine from the task of taking molten glass directly from the tank by causing suitably shaped pieces of glass to issue from the furnace and fall into the moulds of the blowing machine.

The Homer Brooke feeder

The evolution of the gob feeder began before the suction process had been invented by Owens. Homer Brooke emigrated with his family from Yorkshire to the USA in 1849. The family established a mould-making and glass machinery business which Brooke continued on the death of his father in 1863. In 1903 he patented a stream feeder, so called because the glass issued continuously through a hole in the furnace floor. This feeder is shown in Figure 62. Difficulties with the stream feeder arose because the stream of

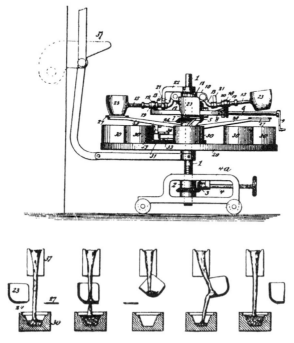

Fig. 62. Homer Brooke's stream feeder. In this early version of the stream feeder the stream of glass from the furnace fills the mould and is then interrupted by a cup and blade. The blade shears off the glass as it moves in the opposite direction to the open sided cup, which partially fills with glass and tilts as it moves under the stream. Meanwhile the next mould has moved into place below the cup and receives the stream as it pours over the edge of the tilted cup. The empty cup rights itself and moves on, the mould fills and the cycle is repeated.

glass continued to flow while the moulds moved intermittently under it. Brooke arranged for a blade and a cup to meet the stream and to contain it, after a mould had been filled, until the next mould moved into place. The cup was fitted with blades which allowed glass in the cup to fall through the bottom when the blades were opened and glass from the cup and the stream then filled the mould, after which the cycle was repeated. The glass stream lost so much heat from its surface that it cooled too rapidly and the finished article contained creases, air blisters and blemishes which were only acceptable for cheap pressed ware. After 1907 several companies made glass jars using Brooke's stream feeder and between 1911 and 1914 the Graham Glass Company, under a licence from Brooke and Louis Proegar who had patented the dividing cup, built a feeder for its semi-automatic bottle-forming machine, the first adaptation of the Brooke feeder to a narrow-neck container machine. This seriously alarmed the Owens Bottle Company who saw the feeder-fed Graham machine as a serious rival to the Owens machine, and they quickly bought up the rights on the Brooke feeder, and soon afterwards the Graham Glass Company itself.

Peiler's gravity feeders

The Owens Bottle Company negotiated licence agreements with other manufacturers but these licences were restricted to specific lines of containers such as milk or whisky bottles. This restrictive policy allowed the number of semi-automatic machines to grow and encouraged the search for new ways of bottle-making which would avoid the necessity of negotiations with the Owens company. An important user at this time was the Beech-Nut Packing Company of Canajoharie, N.Y., who had experienced difficulties in obtaining jars with a finish satisfactory for vacuum sealing. W. H. Honiss and W. A. Lorenz, mechanical engineers of Hartford, Connecticut, and patent attorneys advised the company to build a new glassworks and offered to design new bottle-making machines. Accordingly the Monongah Glass Company was established in 1904 at Fairmont, West Virginia. An Owens machine licence could not be obtained and a new company was formed to design and build the glassworking machinery. The new company was incorporated in 1912 as the Hartford-Fairmont Company, a name which has since become pre-eminent in the glass container industry.

In 1911 the task of designing a new feeder was given to Karl E. Peiler, a graduate engineer from the Massachusetts Institute of Technology. Peiler went to the hand gatherer for inspiration, as is immediately apparent from his punty feeder, but he also designed a paddle feeder to produce gobs of about the same shape and viscosity as the hand gatherer, in contrast to the higher temperature and lower viscosity used in the stream feeder. Peiler's punty feeder is shown in Figure 63.

The punty feeder was not a success. One of his colleagues described it in 'Peiler's Lament':

> My punty lies over in Glenshaw,
> My punty ain't done so damn well,
> My punty is sick at the stomach,
> And W. . . . . . . is chuckling like hell.
>
> Foul curses on the power line
> That fluctuated so,
> And made my punty feeder feed
> Not handsome gobs which we did need,
> But chicken entrails, crude indeed,
> A truly cruel blow.

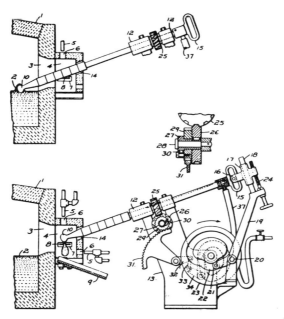

Fig. 63. Peiler's punty feeder. The action of the hand gatherer in taking glass from the glass surface was imitated in the punty feeder. In the hand process the gatherer took up molten glass on the end of the pontil rod by continuously revolving the rod with its end just below the glass surface; the punty feeder was moved and rotated mechanically as shown in the diagram.

In the paddle feeder the glass from the furnace passed through a channel connecting furnace and feeder mechanism, known as the forehearth, where it could be maintained at the correct temperature for the forming process. At the machine end of the forehearth a fireclay impeller, or paddle, worked backwards and forwards in the glass immediately behind a fireclay spout at the end of the forehearth, causing the glass to well up over the spout in a series of waves. Each wave of glass then remained as a suspended mass hanging from the spout until it was cut off by shears and allowed to drop through an orifice ring below the spout into the mould. The quantity of glass urged over the spout could be controlled by altering the effective length of the paddle, the depth to which it dipped into the glass and its distance behind the spout. The Hartford paddle feeder was first used in 1915 by the Monongah Glass Company in the manufacture of jars and it became very popular as a feeder for the Hartford twin-table milk bottle machine. The pear-shaped gobs slid down two fixed chutes on a cushion of steam produced by a fine water spray and were guided into moulds mounted on the twin-tables by a trough swinging between the chutes. Thirty paddle feeders were later installed by various companies and were used in conjunction with the Hartford milk bottle machine and the Hartford press to make disc-mouth milk bottles and pressed ware such as tumblers.

The gobs delivered by these feeders were rather roughly shaped and Peiler sought to produce controlled shapes which would each suit a different finished article. His paddle-needle feeder (P-N feeder) solved the problem. The glass was paddled as before over the lip of the spout and sheared off, but instead of then being allowed to fall freely under gravity through a simple orifice ring it was controlled by the vertical action of a thin refractory plunger or needle which reciprocated above a bowl located below the lip of the spout into which the severed gob dropped. The needle, which varied in shape according to the type of ware being produced, moved downwards into the bowl pushing out a gob of predetermined shape through an orifice in the bottom of the bowl. As with Peiler's paddle feeder, the weight of the gob was controlled by the movements of the paddle over the surface of the glass.

The precise control of weight and shape for the P-N feeder was only possible for high feeder speeds and Peiler therefore set to work on a suitable plunger mechanism for gob formation over a range of feeder speeds, which dispensed with the paddle because the plunger worked within the glass itself as it flowed through a hole in the floor of the extension to the glass tank known as the forehearth which acted as a reservoir of glass. Peiler's Hartford single feeder was introduced in 1922 and in modified forms is used nowadays in glassworks manufacturing containers throughout the world. It is shown in Figure 64. The downward motion of the plunger accelerates the rate of flow; as the plunger is raised the stream of glass thins, the gob is sheared and then falls.

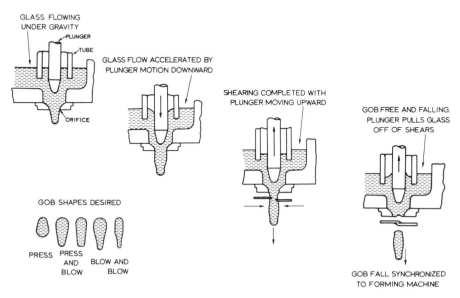

Fig. 64. The single gob feeder. How a gob of glass of the required shape and size is produced for the glass-forming machine.

The Hartford Empire Company offered to sell their feeder and patent rights to the Owens Bottle Company but Mike Owens advised against acceptance, a serious mistake because during the 1920s the Owens Company used increasing numbers of Hartford feeders. In Britain, on the other hand, W. A. Bailey, an importer of high-quality colourless glass containers from the Netherlands took the risk of purchasing the British and Netherlands patents on the Hartford paddle feeder in 1917. In the same year he formed the British Hartford-Fairmont Syndicate to exploit the British rights and the first feeding and forming machines were brought to England for the production of pressed-and-blown ware. The paddle feeder was soon replaced by the P-N type and later by a Hartford single feeder.

In order that these feeders can operate successfully they must work in a constant supply of glass at the correct temperature. The feeder must also drop its gob in such a way that the gob can be fed to the machine. As a result the working end of the furnace has several feeder channels protruding from it. The length of these channels varies between about eight and twenty-six feet and the overall width is about four feet, and the depth about one and a half feet. The temperature of the gob is different for different sizes and shapes of gob but averages around 1120°C for the usual soda-lime-silica bottle glass.

The adaptation of forming machines to work with gob feeders

Until 1917 the Owens machine was the only successful, completely mechanized method of making glass containers. There were, however, several machines in use in which the bottle-making process was partly mechanized. Improvements in these semi-automatic machines included increasing the number of moulds on each machine, the use of electric power to drive and to improve the synchronization of various movements, mechanical devices such as automatic shears, and automatic transfer from the parison to the blow mould.

Eventually, from the 'three-man, two-boy' through the 'one-man, one-boy', a 'no-boy' machine was designed. This machine required only the services of a gatherer or three gatherers to two machines. No-boy machines were produced by the Lynch Glass Machinery Company and the O'Neill Machine Company in 1917. With the advent of the feeder at about the same time the stage was set for the introduction of strong competitors to the Owens machine. The survivor of this evolutionary process has been the mechanical feeder coupled with a new machine developed in the 1920s.

The individual section machine

The machine which has largely displaced its rivals in container manufacture is the individual section, or I.S. machine, invented by Henry W. Ingle of the Hartford Empire Company in 1925. In the I.S. machine the gobs of glass

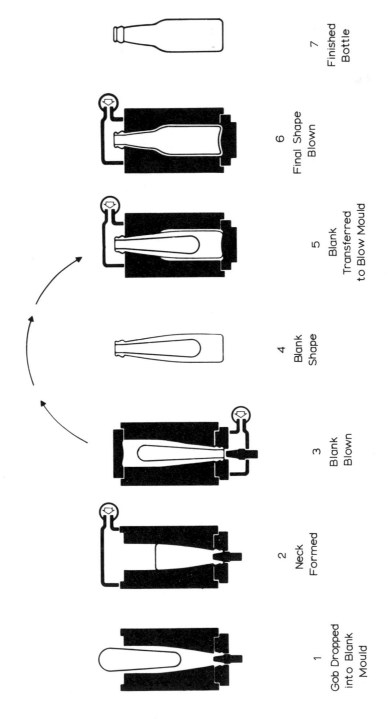

Fig. 65. The automatic blow-and-blow process.

from the feeder are conveyed by chutes to the stationary parison moulds, in contrast to the Owens machine where the moulds are conveyed on a massive moving table to meet the glass. Thus the I.S. machine can be rigidly mounted and the removal of many moving parts makes for easier maintenance. The output per mould is higher because one parison is being formed while the previous parison is being blown in the blow mould. In the Owens machine there are periods of idle time for the moulds between each operation. The I.S. machine was originally developed as a blow-and-blow machine but it can now also be operated as a press-and-blow machine. These two operations are shown diagrammatically in Figures 65 and 66.

The I.S. machine originally consisted of four individual sections, mounted in a straight line on a bedplate, and a conveyor. Nowadays the six-section machine is most commonly used but it can be operated with any number of sections whilst one or more are removed for renewal or repair, thus making it a highly flexible machine. The latest development is the eight-section machine shown in Figure 67 which is capable of production speeds of over 150 bottles per minute.

Container manufacture today

Although a few suction machines are still in use, over ninety per cent of modern containers are made by the gob-feeding system. The rapid spread of this system is illustrated by the history of Beatson Clark and Company of Rotherham and Barnsley, a very successful manufacturer of high-quality containers. A picture of the latest addition to their plant was described in Chapter 5. Hand work persisted until the 1920s and the use of the pot furnace exclusively until 1927 when the first tank furnace was built. Monish suction machines were installed at the end of 1929; at that time ninety-eight per cent of the company's output was mouth blown, the rest being made on semi-automatic press-and-blow machines. Twenty years later eighty per cent of the production was fully automatic, nineteen per cent semi-automatic and less than one per cent mouth blown.

Productivity has been increased by the use of the double- or treble-gob process whereby two or three containers are formed simultaneously in double- or treble-cavity moulds. The double-gob process was patented in the United States in 1939 and was introduced into Britain after the Second World War. Great improvements have also been made in container design, as well as in the forming processes. Hand-blown bottles were thick and heavy because the blower could not control the glass distribution. When automatic machinery was introduced, it was possible to control the glass distribution more easily. After the repeal of Prohibition in 1932 there was an immediate demand for large supplies of beer bottles in the USA and improved designs were introduced. The depression of 1929-31 also created a great demand for good preserving containers for domestic use. It was at this stage that it was

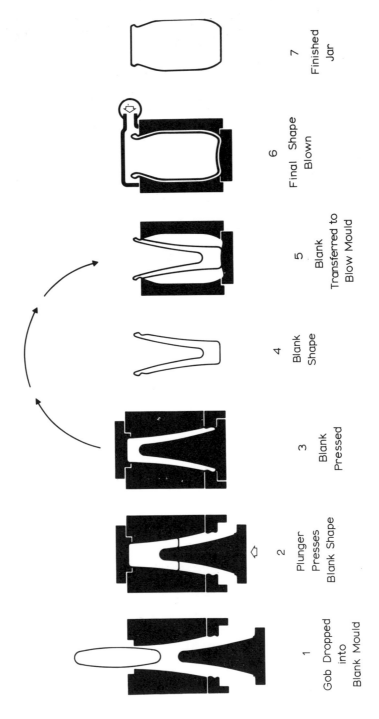

Fig. 66. The automatic press-and-blow process.

Fig. 67. The I.S. eight-section machine. This latest addition to the series of I.S. machines is capable of production speeds of over 150 bottles per minute. The bottles are carried away from the machine on a conveyor belt towards the annealing lehr.

noted that most breakages in the Mason jar originated in a band which extended from the base of the container to about half an inch up the side-walls. The jar originally had parallel sides with sharp corners at the base; it was re-designed with the parallel sides joining the base in a gentler curve. The removal of the sharp corner curve increased mechanical strength and improved the glass distribution. This concept spread to the design of many other glass containers and improvements in techniques of working with smaller amounts of glass led to further weight reductions. In Britain light-weighting of containers was first supported in 1945 by the Glass Container Research Committee of the Glass Manufacturers' Federation which examined all standard designs and re-designed the containers so that they could be made from a smaller amount of glass without impairing their strength or altering the required capacities.

Other methods of improving the strength have been devised. Theoretically, the tensile strength of glass is around one million pounds per square inch, in practice the surface of the glass is abraded and the tiny cracks act as stress magnifiers to any blow on the surface, just like a notch in a bent stick. Thus practical strengths are considerably lower than theoretical strengths and range from 5000 to 10,000 pounds per square inch. The industries using glass containers have installed machines for filling the containers at very high speeds and any breakage can cause a very costly stoppage; any method of improving the practical strength of containers could become of great importance. Silicones and other lubricants give temporary protection against abrasion, but permanent resistance can now be produced by treating containers with metallic oxides such as titanium and tin almost at the moment of forming whilst the glass is still molten. Surface damage by contact is minimized and the friction between the glass surfaces is reduced.

Bibliography

1. *Old English glass: The development of the bottle,* F. Buckley, *Glass,* 1931, 8, 322.
2. *Hollow and specialty glass: background and challenge,* H. Holscher, *Glass Ind.,* 1965, **46,** June-November.
3. *From sand-core to automation: A history of glass containers,* V. Wyatt, A Glass Manufacturers' Federation publication, 1966.
4. *The Ashley bottle machine: an historical note,* S. English, *J. Soc. Glass Technol.,* 1923, 7, 324.
5. *The early development of bottle making machines in Europe,* W. E. S. Turner, *J. Soc. Glass Technol.,* 1938, **22,** 250.
6. *Introduction to the symposium on machinery for the fabrication of glass containers,* F. W. Hodkin, *J. Soc. Glass Technol.,* 1953, 37, 27T.
7. *The introduction of the Owens machine into Europe,* R. S. Biram, *J. Soc. Glass Technol.,* 1958, **42,** 19N.
8. *Automatic glass feeding devices,* G. Dowse and E. Meigh, *J. Soc. Glass Technol.,* 1921, **5,** 134T.
9. *The development of the automatic glass bottle machine,* E. Meigh, *Glass Technol.,* 1960, 1(1), 25.
10. *Revolution in glassmaking,* W. C. Scoville, Harvard University Press, 1948.
11. *Glass Machines,* W. Giegerich and W. Trier (eds.), Springer-Verlag, 1969.
12. *Die Glasfabrikation,* R. Dralle and G. Keppeler, R. Oldenbourg, 1931.

8

Science and the glass industry

In its beginnings glass technology thrived on accidental discovery and the inheritance of accumulated experience. It was influenced by political and economic forces, by government interference in taxation, the granting of monopolies, the use of Royal prerogative. As any other human activity it was a part of the whole structure of society which cannot adequately be viewed in isolation. The very rapid changes of the last one hundred years offer the temptation to divide the history of the technology into pre- and post-scientific eras. To do so, however, would be to ignore much of the content of this book.

For the admirer of the craft of glassmaking, the creation of the Lycurgus Cup, the Portland Vase, the stained glass windows of the Middle Ages, the tableware of the eighteenth century, and modern creative work in glass, have an essential continuity. Nor is there any obvious period of a few years in which a particular phase of the development of the technology can be said to be concentrated. Furnace design, as we have shown, took a major step forward when the Siemens brothers introduced their regenerative furnace but this step could only be taken after a hundred years of scientific endeavour eventually produced an understanding of the nature of heat. The full potential of the new furnace was not realized until, on the one hand, advances in chemical knowledge and the chemical industry produced reliable batch materials and, on the other hand, fifty years of the science of thermodynamics was married with the study of the structure of crystals by X-ray diffraction which, together with the advance of electrical technology, opened the way to the manufacture of greatly improved refractory materials. These advances, together with discoveries in the measurement and control of temperature, made possible the new mechanical methods of glass forming which eventually solved the problems presented by the political and economic pressures of shortage of manpower and demands for high wages. The new mechanical methods were able to satisfy new industrial demands: flat glass for the motor car industry; lamp bulbs for the electrical industry. The result of all these endeavours has been a new and much larger glass industry which has grown from the old craft, leaving the hand craft to be admired and fostered by the prosperity which the new industrial society has made possible.

These changes, which have taken place largely during the last hundred years, have made great social advances possible. Working hours have been reduced, child labour has been eliminated, working conditions have changed remarkably for the better, and many new jobs have been created. These changes are the real achievements of technology.

In a sense glass technology still grows by discovery and the inheritance of accumulated experience and it must remain and change as a part of society. Discovery and inheritance of experience still apply in the sense that discovery is sought energetically and the inherited experience is summarized into systematic knowledge by the methods of science. But much is spoken and written in these days of the impact of science upon industry and in this last chapter we shall concentrate on the interaction between science and the glass industry. In doing so the immense growth of educational opportunities in science and the very large increase in the number of professional scientists employed by the industry must be remembered. We can only summarize by describing the growth of industrial research laboratories and referring to the fact that in 1920 a works chemist was considered a luxury by many small firms, and that physicists at that time were just beginning to be employed by some large industrial organizations.

Industrial research in the twentieth century

Early in the twentieth century large units in the industry were starting to create laboratories where the power of scientific methods and apparatus could be brought to bear upon the problems of industry, upon improvements in the methods of manufacture and in the design of new glasses to meet new needs. The Corning Company of New York State first organized its laboratories in 1908 under the direction of Dr E. C. Sullivan. The first laboratory consisted of one small room and Dr Sullivan had one assistant; by 1961 there were more than five hundred scientists and engineers employed in the research centre.

By the early 1920s the large electrical firms in Britain, on the Continent, and in the United States had laboratories in which a certain amount of effort was devoted to the study of glasses with special properties required in the manufacture of lamps. The first glass problem to arise was solved by the use of what might be called a classical soda-lime glass combined with lead crystal or flint glass. When the first attempts were made to produce incandescent filament lamps entirely of soda-lime glass, it was found that between the wires which led the current to the filament some deterioration of the glass occurred in the form of blackening and bubbling and subsequent development of cracks along the wires; this failure was found to be due to the passage of electricity between these wires. It was soon realized that the electrical conductivity of soda-lime glass was some million times greater than that of lead-crystal glass at the same temperature and so although the cheaper soda-lime glass continued to be used for the bulb, the tube through which the wires were carried to the filament was made of lead glass. It was therefore necessary for the two glasses to have compatible properties so that when they were sealed together the contractions from the temperature at which they became solid down to room temperature were near enough the same to prevent the setting up of strains and consequent cracking. Scientific control was introduced in the form of measurement of the thermal expansion and by studying with polarized light the stresses set up in sealing the two glasses together. When it was found that much more efficient light sources could be obtained by discharging electricity through gases than from a heated metal wire it became necessary to develop glasses which could withstand the attack of the vapour of metals such as sodium and mercury.

From the Corning laboratory came, in 1915, the low expansion Pyrex borosilicate glass which created large markets for domestic cooking ware and also in the construction of large scale chemical plants. This same laboratory began the development of fusion cast refractories. The Company was associated with the birth of a new industry in forming the Owens-Corning Fiberglas Corporation in 1938 for the large scale production of glass fibre, used mostly in thermal insulation and also in electrical applications. The laboratory has long had an interest in glasses which respond to the action of light; photo-sensitive glasses were produced which were analagous to photographic papers and films in that they could be exposed to ultra-violet light through a negative and a positive produced in the glass, not by chemical development but by an appropriate heat treatment. The latest development in this line is a glass containing very small crystals of silver compounds which darkens quite quickly on exposure to sunlight and which recovers its transparency when the intensity of the sunlight falls. The response of the glass to the brightness of illumination is sufficiently rapid for it to be used in sunglasses which automatically darken as the sunlight increases. Unfortunately this glass is too costly for it to be used in glazing windows but the search continues in many laboratories for a cheaper solution.

In this search a lithium silicate glass containing silver was found to crystallize readily when heat-treated after exposure to ultra-violet light, and the crystallized portions of this glass were about 1000 times more soluble in hydrofluoric acid than the parent glass. Thus a glass called Fotoform was discovered which can be made to assume very complex shapes by being exposed to the details of the shape in ultra-violet light; after heat treatment and crystallization the irradiated areas can be removed by hydrofluoric acid.

The age of accidental discovery is not over. One night in 1957 an exposed plate of Fotoform glass was left overnight in an oven for heat treatment, but the temperature controller failed and the oven over-heated by 300°C. The laboratory worker expected to find molten glass in the oven but instead the plate had been converted into a crystalline ceramic. Thus a new class of materials was born which can be prepared and shaped like glassware and then converted into a ceramic by heat treatment. Some of these materials have almost zero thermal expansion and have been marketed in many countries as 'pyro-ceram' domestic cooking ware. Another important application has been in heat-exchangers in gas turbine engines. Powdered glass of suitable composition is coated onto a flexible material which can then be corrugated and interspersed with a coated flat flexible sheet to form a honeycomb structure. The flexible material can then be burnt out and the glass structure remaining converted into a ceramic which can operate continuously at temperatures greater than $600°C$.

Although many of these developments could perhaps have come from painstaking observation of the changes in the properties of glasses with systematic changes in composition, they have also taken place in the context of an ever-expanding development of physics and chemistry, both in the theoretical understanding of the subjects and in the instruments available for making the observations.

Coloured glasses

The introduction of undesirable colour by the presence of iron, the diminution of the iron colour by the addition of manganese, the deep blue of cobalt, the blue-green of copper, the grey colour produced by nickel, and the purple of manganese are all examples of the colours produced by the oxides of these metals when dissolved in glasses. Many of these colouring oxides were familiar even in ancient times but it is only in the last fifteen years that a detailed account of the absorption of light in these coloured glasses has been given in terms of well-defined changes in energy of electrons associated with these ions. The necessary theory was written in 1932 and applied to the optical properties of these ions in the early 1950s. At that time commercial spectrometers were becoming available which produced very quickly and automatically a chart showing the absorption of light and the way in which it varied with the wavelength of the light. Information soon became available about the spectral absorption and therefore, about the colours produced by these ions and the way in which they varied with the state of oxidation and the composition of the host glass. These data could be discussed in terms of energies of electrons and a new foundation was available for the discussion of colour, fluorescence and photo-oxidation, which is the basis of photosensitivity.

The same basis could be used in discussing the colours and fluorescence of rare earths and at this time glasses containing neodymium were shown to be capable of use as lasers. The fundamental principle of the laser was described by Einstein in 1917 and its operation was prophesied by Townes in the 1950s and realized practically in 1960. It is a device by which very intense parallel beams of light can be produced. Lasing materials are mostly crystalline but the glass laser has its own particular applications and the attraction of cheapness and ease of manufacture compared with the crystalline materials which have to be in the form of large single crystals and are very costly to prepare.

The understanding of the optical effects of the metal ions has encouraged attempts now being made to produce very pure glasses which will be so transparent that laser light can be passed down fibres of these glasses with sufficient light left after passing through some miles of glass to enable a system of communications to be built. A system built on these principles could make it possible to send the equivalent of 250,000 telephone messages simultaneously down a fibre less than 0.1 millimetres in diameter.

Science and present technology

Corning Glass Works at one time said that seventy per cent of the products they were selling had not been in production ten years previously. This exemplifies dramatically the possibilities of finding new uses for glasses, sometimes new outlets for traditional glasses and sometimes the production of new glasses with specific properties to meet the needs of other technologies. The established glass industry is also continually responding to scientific knowledge but the time lag between a scientific discovery and its use in industry can be very long. The conditions in industry have to be ripe for new applications. When the Siemens brothers were developing their regenerative furnaces the glass industry was rising out of the trough caused by the taxes imposed on it and a building boom was in progress in the expanding Victorian economy. The need for better refractories was pressing when the equilibria in the alumina-silica system were first investigated, and this new scientific knowledge was quickly assimilated. But the mathematical tools and the physical science were available to solve the problem of annealing glass for at least fifty years before they were so used by Adams and Williamson while helping to develop the American optical glass industry under the pressure of military demands.

Adams' and Williamson's theory, although adequate as a background for the design of annealing lehrs, contained some approximations. When the greatest possible uniformity of properties was sought in large pieces of optical glass it was found necessary to arrange the cooling so that every part of the glass passed through each particular part of the temperature range at the same rate. This specialized piece of technology was a by-product of academic interest in the nature of glasses.

The nature of glasses

In the 1920s great interest arose in the nature of glass in the following way. X-rays had been discovered at the close of the nineteenth century; they were thought to be electromagnetic radiation similar to visible light but of much shorter wavelength. It was also thought, and had been suggested by Ludwig Sieber in 1824, that crystals consisted of layers of atoms arranged in regular patterns. Von Laue in 1912 suggested to two of his students, Friedrich and Knipping, that if indeed X-rays were electromagnetic radiation and the atoms in crystals were arranged in regular patterns, then a narrow beam of X-rays passing through a crystal would be scattered by the atoms but because of the regularity of the atomic arrangement the scattered intensity would be concentrated in certain directions, thus producing a pattern on a photographic plate which would be characteristic of the particular crystal. The pattern was found as predicted and in the hands of W. L. Bragg and his associates this new tool had by 1925 begun to elucidate the structure of silicate crystals.

Glasses on casual inspection are certainly as solid as a typical crystal, but when examined in this way only very diffuse patterns were obtained in place of the well-defined patterns given by the crystals. These diffuse patterns were similar to those given by liquids and, as had been predicted by Tammann in 1925, glasses were shown to be liquids which, although they should have frozen, or become crystalline, at some high temperature, had failed to do so and were therefore super-cooled liquids. But they were a special kind of super-cooled liquid for their viscosity had become so high that they retained their shape without support and had in practical terms become solids. Here was an intriguing situation for the physicist and out of experiments designed to probe into this behaviour came some which showed that the physical properties of a glass, its density or refractive index for example, could be altered by changing the rate at which it was cooled through a range of temperatures around what had been established as the annealing temperature.

Traditionally annealing consisted of cooling glass slowly. This was often done in an oven fired by fuel containing sulphur and it was noted that when the cooling was correct the glass became coated with a bloom which was caused during the cooling by the formation of sodium sulphate from the soda in the glass and the sulphurous atmosphere. This bloom was often the only method of judging whether annealing was satisfactory or not, apart from the fact that the glass had not broken or did not break in the warehouse. Adams' and Williamson's theory of annealing made use of the mathematics of thermal conductivity and of the theory of elasticity, taking account of the fact that in the annealing range, although there can be a temperature gradient in the glass while it is cooling, the glass is soft enough or sufficiently mobile to flow to release any stresses. As it cools down, however, it loses this ability to release stresses and on reaching room temperature the thermal gradient is replaced by a system of stresses. Sufficiently slow cooling would result in stresses so small as to be harmless but even so it is still possible for some parts of the glass to have different refractive indices because it has not all been cooled in exactly the same way. Eventually schedules were designed for annealing large blocks of optical glass which ensure that all parts of the block pass in turn through the same temperature at the same rate thus reducing to a negligible amount the variation of refractive index throughout the block.

At the present time it might be said that this new annealing schedule is the only practical use which has come out of these studies of the nature of glass but this is not entirely true and it may well be that there are larger returns to come. In many ways glasses are convenient for studying some problems of the liquid state and, as has happened before, technology may repay some of its debts to science by aiding in the present active quest for a more complete understanding of liquids.

Problems in the strength of glass

The stresses which are 'frozen-in' a rapidly cooled lump of glass greatly interested Prince Rupert, a nephew of Charles I, and his interest is still recognized in the name, 'Rupert's Drops'. These are made by pouring large drops of glass of the right viscosity into water where they take the form of a pear-shaped drop with a long thin tail. These drops are so highly stressed that if the tail is broken off they fly immediately into millions of very small pieces.

This process was put to technological advantage soon after the motor car came into common use in the development of the toughened windscreen which is made by cooling a sheet of glass very quickly from the annealing range by blowing air onto its surfaces. This leaves the sheet with compression on the outside and tension in the inside. As glass always fails in tension the compression has to be eliminated by an equal and opposite stress before a tensile stress which causes fracture can be built up. Toughened glass is therefore very strong; when it does break the energy stored in the glass on account of the internal stresses causes it to shatter into many small rounded pieces which are not dangerous.

More recently it has been found possible to put the surface layers in very strong compression by a chemical treatment which causes sodium ions to leave the glass and to be replaced by lithium ions. This exchange is achieved by immersing the glass in a molten lithium salt. Typical values of the strength of flat glass are:

Annealed glass	7000 lb/in²
Toughened glass	20,000 lb/in²
Chemically toughened glass	100,000 lb/in²

The science of glass manufacture

The discovery of the nature of heat was followed by the elucidation of the laws of radiation, begun by the theoretical work of Stefan in 1879 and confirmed by the experiments of Lummer and Pringsheim in 1903. Max Planck's interpretation of these discoveries prompted Einstein to lay the foundation for the development of quantum mechanics, an essential precursor to much of today's new technology including solid state electronics, which is having such a tremendous impact on everyday life, particularly in the fields of communications and computers. Radiation pyrometers were developed and the thermocouple was invented at about the same time; by the 1920s they were beginning to be used for measuring and controlling the temperatures of furnaces.

In the 1930s scientific papers on glass discussed the problem of the amount of heat which was able to descend in a glass furnace from the surface of the glass to the bottom of the tank, a distance of about four feet. The bottom of the tank was always found to be very much hotter than it would have been had the heat reached it only by the process of thermal conductivity which occurs around room temperature. This 'normal' process of thermal conductivity arises from an exchange of thermal energy between atoms in an attempt to make the temperature of the whole body the same. It was clear that in the glass tank, that is, in the glass at high temperatures, some extra process was necessary. The other means normally available for the transfer of thermal energy are convection and radiation. By convection is meant the upward movement of hotter parts of the liquid caused by its lower density; it is a process which again tends to equalize the temperature throughout the mass of the liquid. It was clear, however, that because of the high viscosity of glass the convective process could not account for the transfer of heat and, moreover, as the heat was being supplied at the top of the glass, only a very complex convection pattern which arose because of the motion of the glass through the tank could be of any help. Nor was it possible, apparently, to account for the transfer of heat by radiation because the radiation would appear to be absorbed completely in the first inch or so of glass. However, in 1903 Sir Arthur Schuster wrote a paper entitled, *On the transfer of heat from the interior of hot stars* and it turned out that the discussion in Sir Arthur's paper could be used to solve the problem of the transfer of heat through glass. It was shown that radiation reaching a given layer will be absorbed by that layer, thereby heating it and causing it to radiate more energy itself. In this way it was found that although the apparent thermal conductivity of a glass remains at about the room temperature value until the glass has a temperature of about 600°C, it begins to increase rapidly above 600°C approximately at the cube of the temperature on account of the radiative heat transfer. It was not until the 1950s that this particular piece of applied science was introduced into discussions of heat transfer in glass tanks.

The transfer of heat and the change of viscosity with temperature are the fundamental processes in making a glass article. The hand craftsman will rotate the glass on the end of an iron rod to maintain its shape while it cools sufficiently to attain the viscosity appropriate for the next part of the shaping process; he may reheat the partially formed object in a glory hole, an oven-shaped space heated by the furnace, before, for example, putting the handle on the jug or trimming the lip to shape (Figure 68). In the container industry attention is now being given to interpreting the steps in fabrication in terms of heat transfer and flow; in the flat glass industry heat transfer is of

Fig. 68. Automatic glass-forming processes must be carefully designed to ensure that the glass has the correct viscosity at each stage of the process. The hand craftsman is able to exercise this control over his material through long years of experience in working with glass.

particular importance in, for example, the process of toughening glass. In these problems radiative heat transfer can be very important.

These particular examples of the applications of science in the glass industry leave untold the constant development inside and outside the industry of methods of measurement. It is these methods—the very essence of scientific method—that are so essential before control, the currently popular step in industrial development, can take place. The chemist is aided by new physical instruments, automatic spectrographs, flame and atomic absorption spectrophotometers, X-ray fluorescence analysers, infra-red spectrometers, all instruments which have been developed or have advanced tremendously in the last decade. Temperature measurement, devices to indicate the level of the glass in the tank, and measurement of fuel consumption are all used by the plant engineer.

In fifty years the industry has gone through a period of gradually increasing interest in the applications of science and has now reached a period of enthusiastic acceptance of the applications. But now it is passing through a phase where the deployment of scientists in industry is a matter of urgent debate. There are many contributory factors, the changes in educational systems, the increasing complexity of management techniques, the improved quality of the services given by ancillary industries such as scientific instruments and control apparatus. The history of the technology seems to show that the progress occurs only when many factors are propitious; scientific knowledge can only be used when the time is ripe.

Maybe the time is ripe for a period of consolidation. Science may have been too prolific but, if this is so, it is all the more important to keep open good lines of communication with the scientific background. Perhaps the period of consolidation will be one in which the ability to measure so many variables quickly and precisely will be combined with the computer to bring the industry into the age of automation.

Automation and the computer

The increasing use of the computer in the control of industrial processes now makes it necessary to use with caution the words automatic and automation in the glass industry. Automatic and semi-automatic have long been used to describe glass making machinery, semi-automatic meaning that the machine is only operated with manual feeding and unloading. Nowadays the word automation is used when a process is controlled by the collection, at various points on the plant, of data such as temperature or quality of product; these data are fed into a computer which in effect compares them with the desired result and then causes the necessary adjustments to be made. In the chemical industry plants have been so automated that they can gradually adjust themselves to optimize their performance.

Some use of these methods has already been made in the glass industry. In the new float glass plants the glass is cut automatically into various sizes which are decided and adjusted by a computer. These steps can, however, only be taken when the time is ripe; in the container industry it will probably be necessary before much automation is introduced to wait until more progress has been made with mechanizing the inspection process which at present is still almost entirely manual.

History is a matter of collecting information, frustrating perhaps because the collection is always incomplete and the interpretation of the information imperfect. Prophecy is much more difficult. After having attempted so briefly to summarize the impact of science on the industry it is appropriate to end with the following quotation from the address of the President, Sir Cyril Hinshelwood, to the tercentenary meeting of the Royal Society, almost exactly 300 years after the publication of the translation of Neri's book by Merrett, a founder member of the Society:

> The propounding of ambitious practical aims seldom leads to their fulfilment. 'Find me a way', an ancient despot might have said,

'to send invisible messages round the world and I will ennoble and enrich you'. The reward would not have been won. 'Find out', he might less plausibly have said, 'about the strange force in this little piece of amber'. This apparent frivolity contained the secret of his first requirement, but remote and unforeseeable.

Acknowledgements of illustration sources

Copyright of the Trustees of the British Museum: jacket; plates 1, 2, 3, 4, 5, 11, 13, 14; Figures 14 and 27. Drawings by S. Frank: map and Figures 16, 18, 20, 21, 23, 34, 37, 50, 51. Department of Glass Technology, University of Sheffield: plates 6, 7, 10; Figures 1, 3, 4, 6, 10, 15, 25, 30, 31, 32, 33, 41, 42, 43, 44, 45, 46, 56. Courtesy of the French Government Tourist Office: plate 8. Copyright of the Oxford University Press: Figure 2. Courtesy of the Smithsonian Institution: plate 9. Courtesy of the *London Illustrated News*: Figures 5 and 47. Courtesy of the Glass Manufacturers' Federation: Figures 7, 8, 9, 11, 12, 22, 53, 54, 55, 57, 58, 59, 65, 66, 68. Courtesy Owens-Illinois, Toldeo, Ohio: Figures 58 and 59. Courtesy *The Glass Industry,* New York: Figures 60, 61, 64. Courtesy of the Corning Museum of Glass, Corning, New York: plate 12. Courtesy of the Magadi Soda Company Ltd., Winnington, Northwich, Cheshire: Figure 13. Courtesy of the Society of Glass Technology: Figures 17, 62, 63. Courtesy of VEB Carl Zeiss, Jena: Figure 19. Courtesy of Glass Tubes and Components: Figure 29. Courtesy of Joseph McCarthy & Company, and the Newcastle *Evening Chronicle*: Figure 24. Crown copyright, Science Museum, London: Figures 26, 28. The Patent Office, London and Sheffield City Libraries: Figures 35, 36. Courtesy of King, Taudevin & Gregson Ltd., and Rockware Glass Ltd.: Figure 38. Courtesy of King, Taudevin & Gregson Ltd., and Glass Bulbs Ltd.: Figure 39. Courtesy of King, Taudevin & Gregson Ltd., and Beatson, Clark & Co. Ltd.: Figure 40. Courtesy of Pilkington Bros. Ltd., St Helens: Figures 48, 49, 52. Courtesy of United Glass Ltd.: Figure 67.

Table 1 is reproduced from *Ancient Egyptian Materials and Industries* by A. Lucas, published by Edward Arnold Ltd. Table 2 is reproduced by permission of the Society of Glass Technology.

Index